1st Regional Process Technology Conference 2009

Galveston, Texas, USA
1-2 October 2009

ISBN: 978-1-5108-1773-9

Printed from e-media with permission by:

Curran Associates, Inc.
57 Morehouse Lane
Red Hook, NY 12571

Printed by Curran Associates, Inc. (2016)

For permission requests, please contact AIChE
at the address below.

AIChE
120 Wall Street, FL 23
New York, NY 10005-4020

Phone: (800) 242-4363
Fax: (203) 775-5177

www.aiche.org

Additional copies of this publication are available from:

Curran Associates, Inc.
57 Morehouse Lane
Red Hook, NY 12571 USA
Phone: 845-758-0400
Fax: 845-758-2634
Email: curran@proceedings.com
Web: www.proceedings.com

TABLE OF CONTENTS

(1) PHYSICAL PROPERTIES AND PROCESS SIMULATION – SPONSORED BY THE NATIONAL THERMODYNAMIC LABORATORY AT TEXAS A&M UNIVERSITY GALVESTON

(2) PROCESS SAFETY

(3) ADVANCES IN PETROCHEMICALS TECHNOLOGY

KEYNOTE ADDRESS

(4) DISTILLATION AND SEPARATION

(5) ENERGY EFFICIENCY – SPONSORED BY SOTEICA LLC, TLV AND RAWSON

(6) CURRENT ISSUES AND OPPORTUNITIES IN REFINING

2009 Regional AIChE® Process Technology Conference – Galveston, TX – October 1st & 2nd 2009
Session: Physical Properties & Process Simulation

Thermophysical Properties and Phase Equilibria: Best Practices in Process Simulation Software

by:

Ashok K. Dewan, DIPPR® Technical Director

DIPPR® (Design Institute for Physical Properties), under American Institute of Chemical Engineers (AIChE)

October 1st, 2009

Contact: ashok.dewan@shell.com, +1 281 544 7664

AIChE

Presentation Outline

1. **Introduction**
   - Getting Started – A Lesson From History
   - Energy Crisis / Shortage - Demand & Supply
   - Step Change in Energy Supply - Alternative Fuels
   - Incremental Energy Improvements - Pinch Analysis
2. **Improved Process Design**
   - Introduction to Process Engineering
   - Current Landscape for Process Engineering
   - Engineer's Toolbox – Plethora of Tools
   - Process Design and Flowsheet Synthesis
   - Importance of Physical Properties in Process Calculations
3. **Real Life Lessons Learned – Improper Use of Physical Properties**
   - Thermodynamic Property Examples
   - Transport Property Examples
4. **Creating Business Value Through Process Modeling**
   - Value Proposition & Keys for Successful Process Modeling
   - Why Process Modeling Can Fail
   - Opportunities for Process Modeling
   - Process Integration & Pinch Technology
5. **Conclusions**

AIChE

1. Introduction

AIChE

Getting Started - A Lesson From History

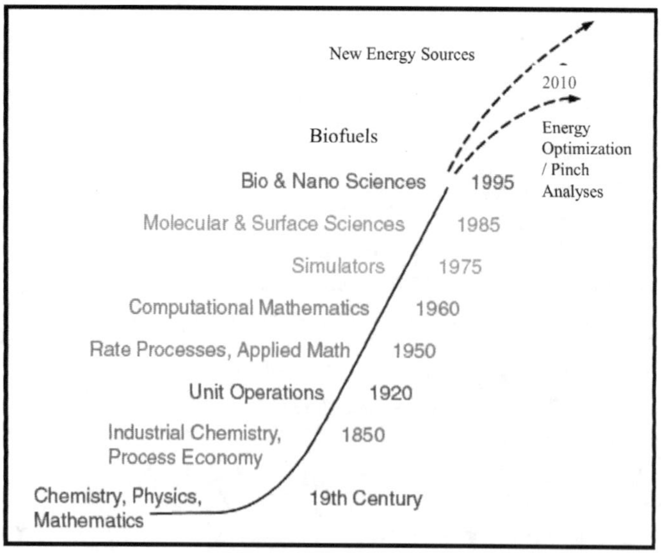

AIChE

Energy Crisis/Shortage - Demand & Supply

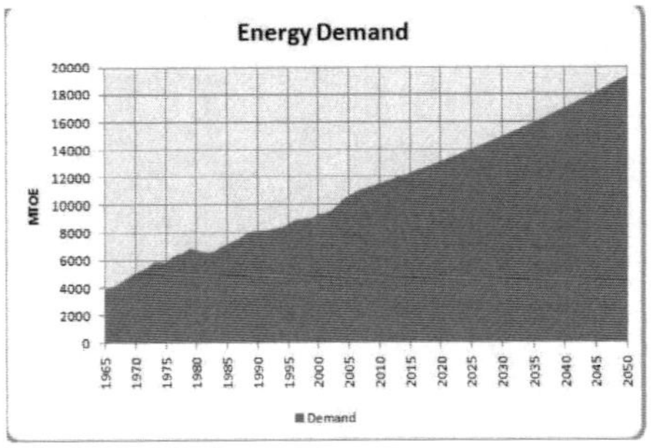

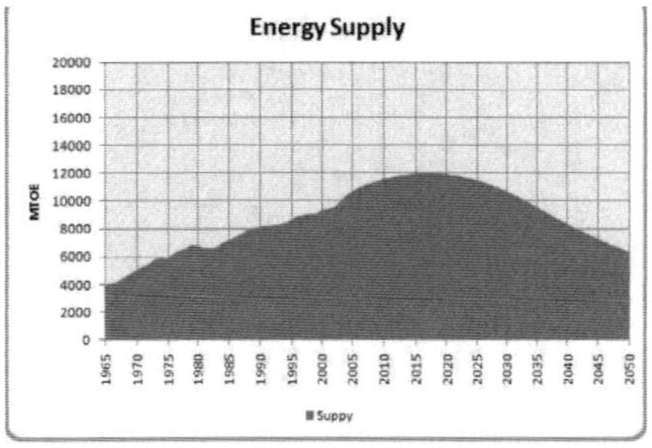

Source: BP Report "Statistical Review of World Energy 2007"

AIChE

Step Change in Energy Supply - Alternative Fuels

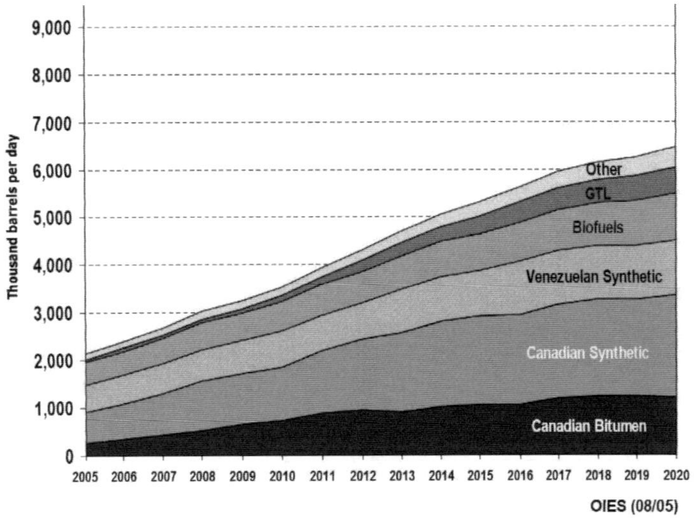

Source: Skinner, R., *Difficult Oil.* Oxford Institute for Energy Studies Sep 2005

AIChE

Incremental Energy Improvements – Pinch Analysis

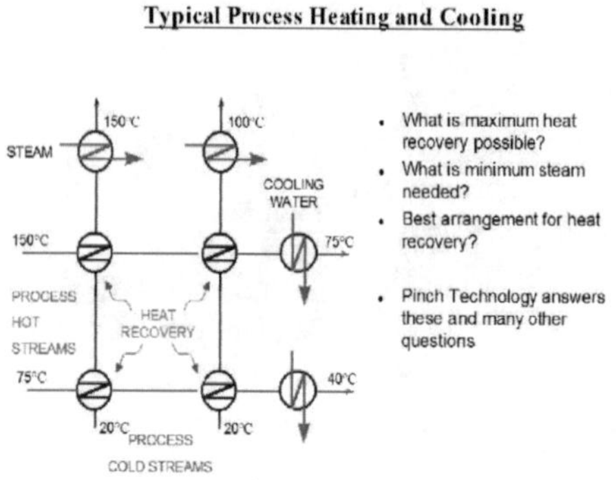

Increasing Energy Efficiency Through Traditional Application of Pinch Technology to Hot/Cold Streams

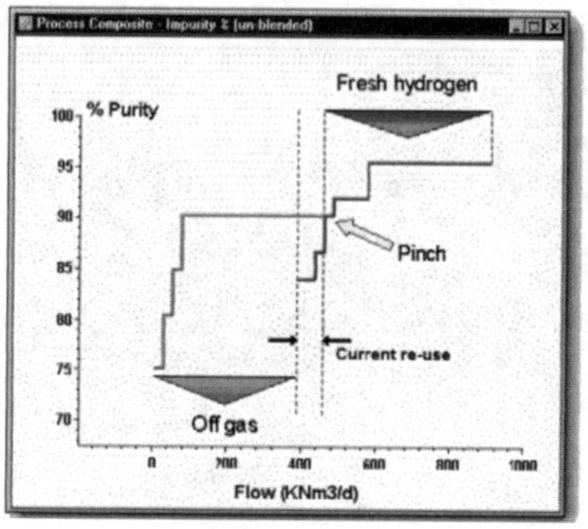

Extending Pinch Technology Applications to Other Refinery Utility Systems (e.g., Hydrogen System, Waste Water Reuse, Greehouse Gases Reduction, etc.)

AIChE

2. Improved Process Design

AIChE

Improved Process Design – Introduction (1)

What Is "<u>Process Engineering</u>" ?

- A fascinating and challenging profession that covers many disciplines
  - Finance, Economics, Engineering (all of them), Management… (you name it!)
- Focused on the design, operation and maintenance of chemical and manufacturing processes
  - Vast range of industries: Oil & gas, chemical, food, paper, power, metal

AIChE

Improved Process Design – Introduction (2)

What Is "**Process Design**" ?

- The specification of physical and/or chemical transformations of raw materials needed to manufacture a desired product
  - Many steps to be selected: reaction, separation, mixing, heating, cooling, pressure change, particle size reduction/enlargement

AIChE

Improved Process Design – Introduction (3)

What Is "**Process Synthesis**" ?

- Interconnecting several transformation processes into a working flowsheet
  - Traditional process synthesis follows the onion diagram

ONION DIAGRAM

Traditional approach:
From the center outward

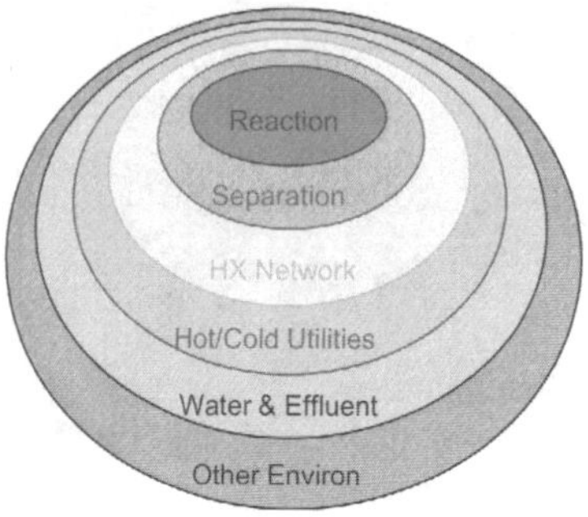

Ideal approach:
All layers together

AIChE

Improved Process Design – Current Landscape for Process Engineering

Support Work Force

- ✓ **Global Team Lead**
- ✓ **Regional Focal Points**
- ✓ **Export Compliance**
- ✓ **Local Expectations**
- ✓ **7/24 Servicing**
- ✓ **"Setting Sun" Paradigm**
- ✓ **Out-Sourcing IT activities**

Tactical Considerations

- One Size Fits All (standard tools)
- Rationalization of Portfolio
- Minimization of Versions
- Model Migrations between Versions
- "Standardized" Hardware
- Harmony with IT Support
- Cannot Please All
- Integrated with Evolving & Changing Business Plans

Playing Field Realities

- License Management & Distribution, on "locked-down" desktops, central Citrix servers, and Standalone – Legacy / Commuting
- Corporate Servicing, across company lines, JV's, Affiliates, etc.
- "Evergreen Communication" via Training, Workshops, Road Shows, Lunch n' Learns, Vendor Liaison, Consortia, etc.
- Gatekeeping / Customizations
- Challenging internal Technical Issues
- Tight License Management thru' Token Rings, temporary "commuting", and "standardized" protocols for installation, scripts, and IT access
- Technical Simulation/Data HELP Desk Support – on demand
- Company customizations needed to add internal thermodynamic and transport properties technology to commercial programs
- Development and maintenance of "Phase behavior guide" and native programs

AIChE

Improved Process Design – Engineer's Toolbox – Plethora of Tools!!!

AspenTech

UNISIM

SIMSCI

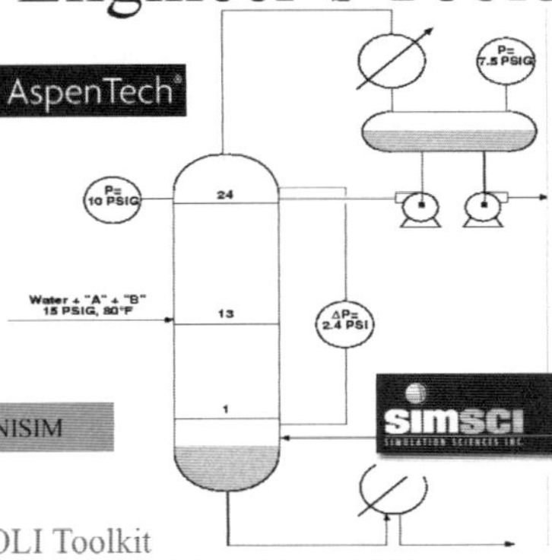

OLI Toolkit

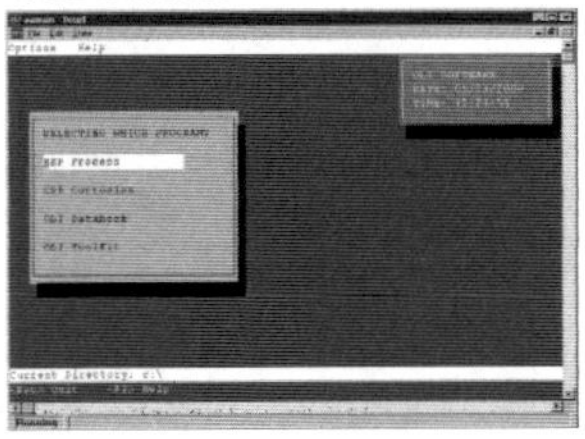

Engineer's Aide - CHEMPRO

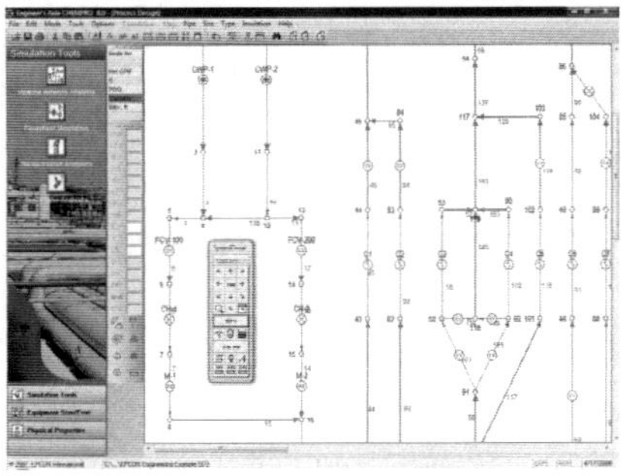

AIChE

Improved Process Design – Role of Process Modeling

Mass / Energy Balances across multiple process units that define a "process"

Key features:

- Raw materials and products yield
- Equipment sizes and operating conditions
- Environmental impact - **(GREEN Engineering & "Safe" Operations)**
- Conditions of all process streams (inlet / outlet / intermediate)
- Economic performance

Evolution Of Process Modeling:

1960's : Sequential Modeler Steady-State Flowsheeting on Mainframes

1970's & 1980's: Commercial Process Simulators – PROCESS®/PRO-II®, AspenPlus®, HYSIM®/HYSYS® (Mostly Steady State, Some Dynamics)

1980's & 1990's : CAD Tools, 3-D Visualizations on Graphics Work-stations; Electrolytes & Polymers; PC-based Process Simulators (e.g., Design-2000®, ProSim®, VMGSim®, …)

1990's & 2000's : Equation-Oriented Solvers & Database Archives; RATE-Based Separations, EXCEL-based computing at PC (Microsoft-centric IT World); Mixed-Solvents

2000's & 2010's : Dynamic Simulation, 3-D Visualizations, Front-End-Engineering-Design, Process Simulation Standards, More "What-If's" To Optimize, Highly constrained, "locked" corporate work-stations ; Virtualization (APP-V)

AIChE

Improved Process Design – Process Design & Flowsheet Sysnthesis

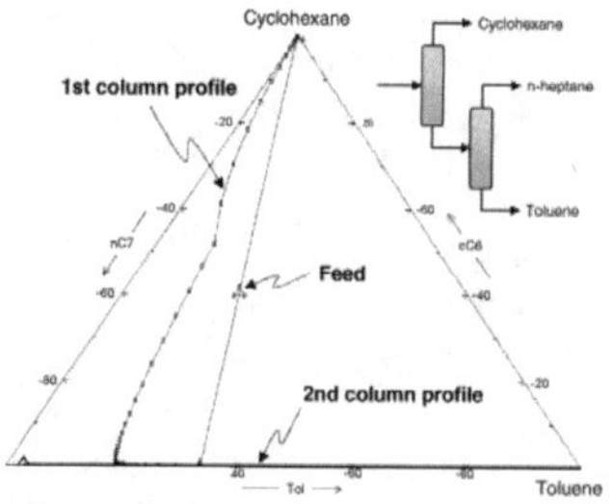

Modern process flowsheeters now provide access to multitude of databank compilations (DIPPR, NIST, DDB), thermodynamic data consistency checks, binary / ternary plots, residual curves, process synthesis tools, 2-way data transfer to MS-EXCEL, heating / cooling curves, azeotropes, gas hydrates, etc.

Many “

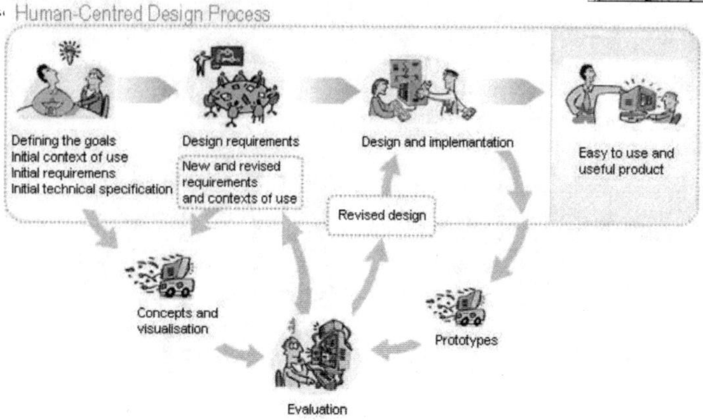

AIChE

Improved Process Design - Physical Props. How To Get Started ?

- Which properties to “sweat” and which to “sweep” (as in, “under the rug”)? What’s needed for job at hand?
- Has this compound (or, mixture) been modeled before?
- Can you lump components, use model compound?
- What am I modeling? -- e.g., distillation column sizing, and pressure-drop calculations -- will require vapor pressure , VLE / LLE, and will additionally require certain transport properties:
    - Liquid and vapor thermal conductivity
    - Liquid and vapor viscosity
    - Surface tension

AIChE

Distillation Column Sizing / Specification

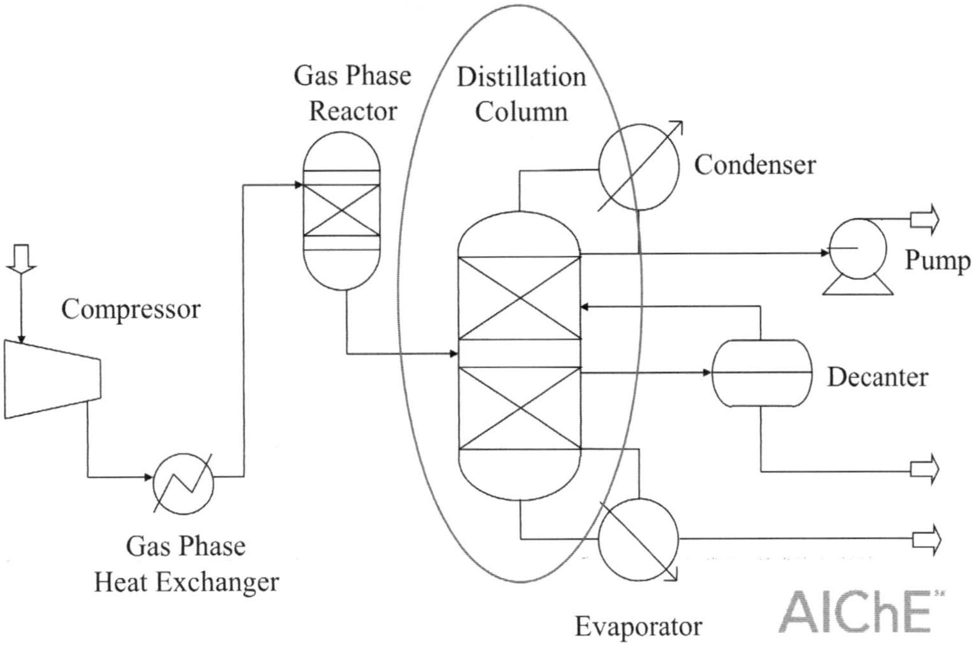

AIChE

Physical Properties For Distillation Column

Equipment Properties	Distillation Column
Thermal	vapor density liquid density
Caloric	liquid heat capacity heat of vaporization vapor pressure
Phase Equilibria	$K_i = y_i / x_i$
Interfacial Properties	surface tension
Transport Properties	liquid viscosity liquid diffusion coefficient vapor diffusion coefficient
Chemical Reaction Rates	Gibbs free energy

AIChE

Physical Properties
What's The Beef ?

THE PROBLEM:

You are responsible for the accuracy of the physical property data you use in your calculations.

But ….. You are rarely given the instruction and tools required to properly find, calculate, or verify your physical property data!

AIChE

Selecting the "Right" Thermodynamic Model

✓ Focus on the following:
- ❖ the physical properties of interest for chosen process units
- ❖ the composition of the mixture (infinite dilution, concentrated)
- ❖ the pressure and temperature range (sub-critical, super-critical)
- ❖ the availability of parameters (cryogenic, vapor association)

✓ Simulators offer predefined thermodynamic models
- ❖ study them well – one method does **not** fit all systems
- ❖ review "mixing" rules – "mixture" properties from "pure"
- ❖ confirm validity ranges – parameters may be limited in coverage of temperature, pressure and composition spans

AIChE

Validating the Underlying Parameters

- ✓ **Ignore the hype**! Process simulator databanks are "**limited**" in physical properties coverage – 2500 compounds, 8000 binaries, selected ternaries, sparse "experimental" data, "narrow" T, P, X spans, and 30+ K-Value methods to whet the appetite!
- ✓ Are the most important pure component and binary/ternary physical properties covered adequately, in T, P, X range of interest?.
- ✓ Make "point" validity checks for the important pure/binary systems with known "good" experimental data (e.g., DIPPR, TRC).
- ✓ For mixtures, use common-sense rules to classify high/high, high/medium, medium/medium, medium/low, low/low binaries. The "key" pairs need experimental data; others can be estimated.
- ✓ Generate tables and graphs to verify "correct" extrapolations.

AIChE

Identifying Non-Library Components and Missing Physical Properties Data

- ✓ If at first you don't succeed in finding your components, try synonyms and formula searches in your favorite flowsheeter.
- ✓ Is this a major component (or, purity spec)?
  - ❖ If neither, can it be lumped or dropped out of the simulation?
- ✓ Does the component (e.g., inert) hold the phase open?
- ✓ Is the component non-volatile?
- ✓ Will reaction (including decomposition) cause this component to be depleted?
- ✓ What properties need to be accurate for my simulation?

AIChE

Obtaining "Missing" Physical Property Data

- ✓ The **recommended** order of data search is:
  - ❖ **1.** Critically evaluated data sources (DIPPR-801/911);
  - ❖ **2.** Non-evaluated sources (TRC Floppy books);
  - ❖ **3.** Experimental measurements (in-house / contract lab)
  - ❖ **4.** Estimation techniques (Joback, UNIFAC, Mol. Model.).
- ✓ Due to the large number of binary pairs in even a simulation of just ten components, rank-ordering the components and binary pairs is a "must" for structured literature search and measurement efforts.

AIChE™

Improved Process Design – HS&E, Equipment & Custody Transfer Properties

Health, Safety & Environment (HSE) Properties:

- ✓ **Fire- and explosion-related properties**
  - ❖ **Flash points, the "fire triangle", minimum oxygen concentration, flammability limits, autoignition temp.**
- ✓ **Henry's law constant in water**
- ✓ **Chemical compatibility**
- ✓ **Dust Hazards**
- ✓ **Impact sensitivity and energetic compounds**
- ✓ **Organic peroxides and peroxide-formers**

Equipment Design / Equipment Sizing Properties:

- Thermodynamic Properties (e.g., Vapor Pressure, K-Value)
- Transport Properties (e.g., Viscosity, Heat Capacity)
- Specification Properties (e.g., Pour Point, Cut Point)
- "Permit" Properties (e.g., Air, Sulfur, Waste Water)

Custody Transfer Properties:

- Thermodynamic Properties (e.g., Liquid Density)
- Transport Properties (e.g., Viscosity)

AIChE™

Improved Process Design – The Seven Essentials For "Good" Physical Properties

✓ **Select** the appropriate thermodynamic model!

✓ **Validate** the underlying databank parameters!

✓ **Characterize** non-library components!

✓ **Identify** all (lib/non-lib) missing physical property data!

✓ **Beg / Borrow / Buy** "missing" physical property data!

✓ **Estimate** some missing physical property data!

✓ **Include** relevant Transport / Interfacial Properties!

AIChE

3. Real Life Lessons Learned – Improper Use of Physical Properties

AIChE

Liquid Vapor Pressure:
A Real-Life Lesson-Learned, No. 1

✓ Very difficult distillation separation
✓ All available v.p. data were at or near 1 atm
✓ Extrapolation suggested sep. easier under vacuum
✓ New measured data: verified easier under vacuum !

Moral of the story:

✓ Determine which properties (for which compounds) are the most important to the situation, and measure if not satisfied

AIChE

Liquid Vapor Pressure:
A Real-Life Lesson-Learned, No. 2

✓ V.P. was extrapolated to 0 °C for environ. permit
✓ Later found to be off by 50%

Moral of the story:

✓ Beware of extrapolations, esp. <40 torr or > 1 atm
✓ Always consider uncertainty vs. consequences
✓ Verify w/ measured data if consequences are important

AIChE

Liquid Vapor Pressure: A Real-Life Lesson-Learned, No. 3

✓ Estimated the NBP of a C_{16} compound using Joback method in a process simulator

✓ Later estimated via spreadsheet by Stein & Brown method – difference of over 100 °C!

Moral of the story:

✓ Estimation of boiling points is often unacceptable even for minor components. Beware of the uncertainties of the estimation methods!

AIChE

Heat of Vaporization: A Real-Life Lesson-Learned No. 1

✓ In-house distillation simulator regresses ΔH_v vs. T to a polynomial equation "on the fly"

✓ Calculated <u>negative</u> values for ΔH_v near and above T_c for an important compound!

Moral of the story:

✓ Keep an eye on T_c

✓ Watch out for regressions and extrapolations, especially with polynomial equations

AIChE

Heat of Vaporization:
A Real-Life Lesson-Learned, No. 2

- ✓ A popular chemical flowsheeter has a known bug of sometimes estimating ΔH_v's which <u>increase</u> with increasing T (wrong trend)
- ✓ This in turn causes the calculation of <u>negative</u> liquid heat capacities

Moral of the story:

- ✓ Always look closely at any and all estimated values, for all properties

AIChE

Liquid Density:
A Real-Life Lesson-Learned, No. 1

- ✓ Effect of pressure on liquid density: very little. Density increases but slightly, and only at very high pressures
- ✓ A dangerous effect: The effect of <u>density</u> on <u>pressure</u>!
  - ❖ "Blocked-in" pressure buildup: Pressure increase of approximately 100 psi for ethylene glycol, 825 psi for water, 1600 psi for acetone, and 2400 psi for methyl acetate

AIChE

Mixture Liquid Densities: A Real-Life Lesson-Learned, No. 2

- ✓ In a simulator, the tabulated pure component liquid densities *might not be used at all when calculating mixture densities!*
- ✓ Example: pure component densities from DIPPR® database and equation, but mixture densities calculated via Rackett method

Moral of the story:

- ✓ With a simulator, watch the mixture densities as well as pure component densities

AIChE™

Liquid Density: A Real-Life Lesson-Learned, No. 3

- ✓ Beware: Process simulators' liquid density estimations will sometimes yield wildly inappropriate values

Moral of the story:

- ✓ With a simulator, choose your estimation method wisely, and critically evaluate the results using the rules of thumb and typical values given earlier

AIChE™

Heat of Formation:
A Real-Life Lesson-Learned

✓ For chemical reactions, must make sure the ΔH^o_f values are "consistent" with each other.

✓ If they are "inconsistent", the calculated heat of reaction is likely to be greatly affected by the uncertainties in the ΔH^o_f values.

✓ In fact, the uncertainties could very well be as large as, or even larger than, the true heat of reaction!

AIChE

Heat of Formation:
A Real-Life Lesson-Learned

Liquid-Phase Standard State Heats of Formation for 2-Hexanol and 3-Hexanol

	ΔH^o_f (25°C) (kJ/gmol)	
Compound	**Literature**	**Estimated**
2-hexanol	-388.40	-393.69
3-hexanol	-392.40	-391.51

(Data source: E. S. Domalski and E. D. Hearing, *J. Phys. Chem. Ref. Data*, Vol. 22, No. 4, 1993, pp. 805-1159.)

AIChE

Liquid Viscosity:
A Real-Life Lesson-Learned

✓ A large heat interchanger was designed based on a certain component for which η_L was extrapolated to be in the 10-25 cP range. The viscosity was later found to be significantly higher, and the interchanger was 20-30% undersized.

Moral of the story:

✓ If one particular property of one particular compound or mixture is critically important to a process design, make sure it's right!

AIChE

4. Creating Business Value Through Process Modeling

AIChE

"Value Proposition" for Successful Process Modeling
- Boost Productivity -

Value Proposition:

- "**Productivity**" is directly proportional to "**Results Obtained**" and inversely proportional to "**Resources Consumed**" :

  Productivity = (Results Obtained) / (Resources Consumed)

- **Increase Results Through:**
  * Better distillation models that tighten purity specifications
  * Better capacity models to increase throughputs
  * Better kinetics models to maximize product yields
  * Better instrumentation and APC to reduce off-spec operations
- **Decrease Resources Consumed Through:**
  * Better utilization of Physical Property external sources – NIST, API, DDB, GPA, Knovel on-line service – to minimize experimentation
  * Pinch Analyses studies to cut energy consumption
  * Hydraulic modeling to optimize piping network and remove bottlenecks
  * Production allocation models to reduce total operating costs
  * Supply-chain models to minimize inventory

AIChE

Keys for Successful Process Modeling

- Define l**ong-term goals** (document, maintain, integrate into work flow, and measure productivity) & **retain repositories** (custodian, update models to newer editions of tools/simulators, replace as needed) for Process Models.

- Provide **process modeling resources** (create new models, train, bug fix, manage versions of toolkit).

- Develop a **robust portfolio of modeling tools** (data tools, flowsheeting tools, process control tools, CFD tools, hydraulics tools, system analyses tools, logistics tools, finite element analyses tools, ...).

- Maintain the **IT infrastructure** (locked-down PC's, servers, company IT protocols, deployment documentation, ...).

- Align **with the Business** (customer intimacy, user groups network, ...)

AIChE

Why Process Modeling Can Fail

- Victim of **becoming common-place** – taken for granted that models are available for everything, under all conditions and for all types of process customizations – any one can use them and crank results.
- **Unrealistic customer expectations** – any computer tool can be used for any type of customer problem – use customer dictated choice of tool instead of optimum tool.
- **Model Overkill** – one size tool fits all problems, ignoring simpler ways to solve problem.
- **Using Improper Tool** – oversimplifying customer problem to fit within constraints of computer tool, instead of choosing right tool for task.
- **Johnny-come-late syndrome** – when all else fails invoke a tool at eleventh hour to solve a complex customer problem, taking higher risks than warranted.
- **Ignoring the end-customer** – delivering process models with limited documentation, no explanation of pitfalls if used outside current conditions, no customer participation in solution development, and no training in computer tools portfolio used
- **Developing non-standard IT solutions** – using sophisticated computer-intensive non-standard IT platforms (e.g., UNIX cluster for CFD problems) to solve customer problems, with customer barred from using the computer tools.
- **No follow-up services** – customers left to themselves to interpret / use complex computer tools and process models with built-in traps.

AIChE

Process Integration and Pinch Technology

Process Integration (PI):

- A holistic approach to process design and optimization, which exploits the interactions between different units in order to employ resources effectively and minimize costs
  - How to reduce Opex-Capex by exploiting the interactions among process units
  - Better results by integrating than by analizing individual units separately

Pinch Technology:

- PI tools that make use of quantity vs. quality charts to represent and set targets for resources demands (energy, water, etc) in complex systems
  - Tools to visualize/account for interactions of many processes added together
  - Once interactions have been considered, true resources targets can be set

AIChE

Work Flow for Improving a Process Plant

- Pinch Technology provides most of the PI tools for process design
  - PI-Pinch Technology is complementary to simulation / modeling tools
- PI-Pinch provides insight and guidance to better integrate processes
  - A step before or in between process simulation & evaluation
  - Applied during plant design, retrofit, and/or operational optimization

Cycle for Improving/Integrating Complex Process Systems

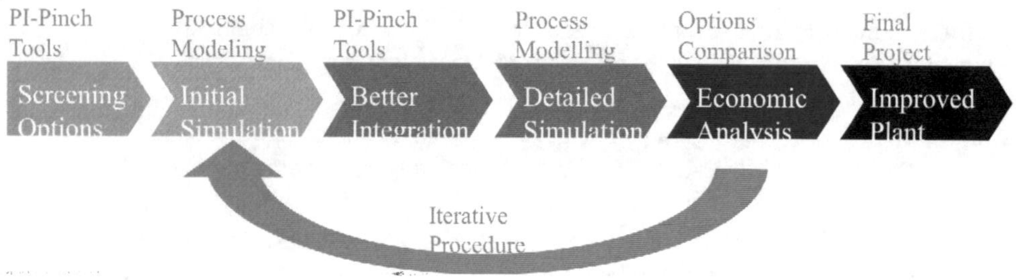

AIChE

Pinch Technology Applications

Pinch Technology was originally developed in the 1980s to reduce energy consumption, since then, it has evolved into many other areas:

Improvement Projects	Typical Resources	Main Objectives	Some Examples
- Plant Design	- Energy	- Reduce Opex / Capex	- Decrease energy input
- Plant Retrofit	- Water	- Minimize resource use	- Emissions abatement
- Plant Operation	- Hydrogen	- Minimize waste reject	- Utilities allocation
	- Feedstocks	- Optim Capex allocation	- Optim batch processes
		- Debottlenecking	- Refinery scheduling
		- Meet product specs	- Reactor sys design
			- Separation sequencing
			- Wastewater reduction
			- Hydrogen production

AIChE

Pinch Technology Applications

- Typical industries in which Pinch Tools have been successfully applied include:

- Oil and Gas	-Food and drink
- Chemicals	- Steel and metal
- Pulp and paper	- Power generation

Decreasing Resources Consumption – Potential Savings

- Application of Pinch Tools in these sectors can typically identify*:

  - Savings in energy consumption: 10-35%
  - Savings in water consumption: 25-40%
  - Savings in hydrogen consumption: up to 20%

Main strength of PI-Pinch is to exploit interactions that cannot be appreciated by analyzing each process separately!

*Source: Pinch Analysis for the efficient use of energy, water & hydrogen. Report by the agency of Natural Resources Canada, 2003. www.cetc-varennes.nrcan.gc.ca

AIChE

6. Conclusions

AIChE

Conclusions

- Process Modeling using an appropriate Process Engineering Toolkit can improve Productivity significantly
- Process Models built earlier in the product development cycle, with full participation of end-customers, will produce low-cost, robust facilities
- Plant designs today are subject to tougher constraints and requirements
- Process Modeling can be conducted at various levels in the organization – plant, ... process, ... unit operation, ... equipment, ... instrumentation, ... logistics, ... APC, ... operator training, ... "greener" operations
- Physical properties, ... chemistry & kinetics of reactions, ... thermodynamics of phase equilibria, ... corrosion rates, ... air/water permits, ... set point targets, ... process workflow & dynamics, ... Process Modeling is Multi-Disciplinary and Integrated Process Models Provide "Best" Solutions
- A penny saved is a penny earned - Pinch analysis can be used to improve existing heat exchanger network designs:
  - Potential energy savings (~15% average) is very possible
- Process Modeling and robust P. E. Tools can be used to overcome new challenges – unconventional oil, ... biofuels, ... new materials!

AIChE

ACKNOWLEDGEMENTS

Special thanks to :

AIChE® STS for organizing this session, and to

DIPPR ®, especially Jeff Smith, Chair of DIPPR Project 801, who compiled the "real-life lessons learnt" physical property examples, for an AIChE Continuing Education course in Physical Properties

Contact: ashok.dewan@shell.com, +1 281 544 7664

AIChE

Thermophysical Properties Research at Texas A&M

Kenneth R. Hall

Artie McFerrin Department of Chemical Engineering
Texas A&M University
College Station, TX 77843-3122

Artie McFerrin Department of CHEMICAL ENGINEERING | TEXAS A&M ENGINEERING

TEXAS A&M UNIVERSITY *at* QATAR

ISOCHORIC APPARATUS

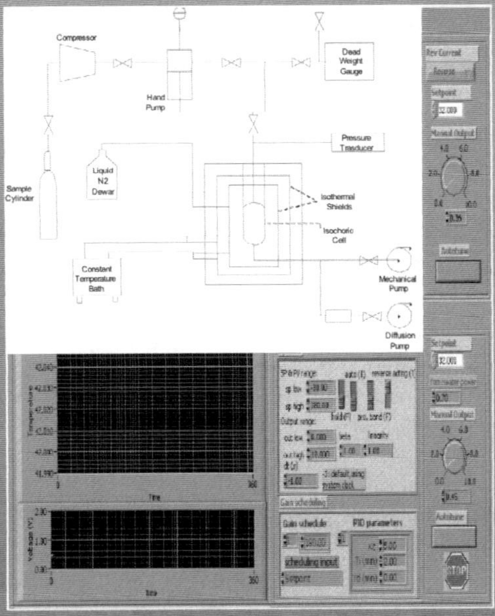

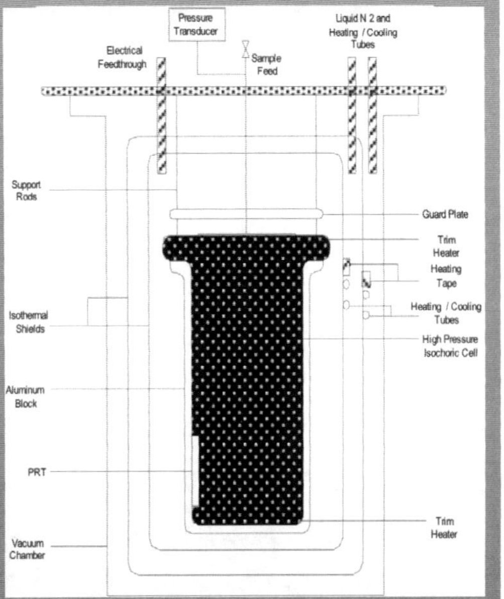

ISOCHORIC APPARATUS

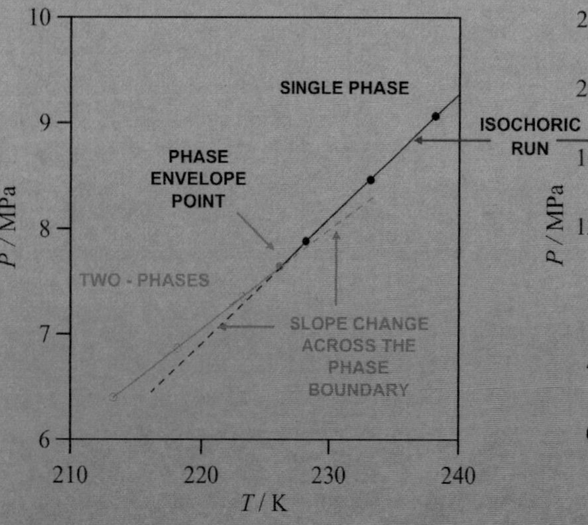

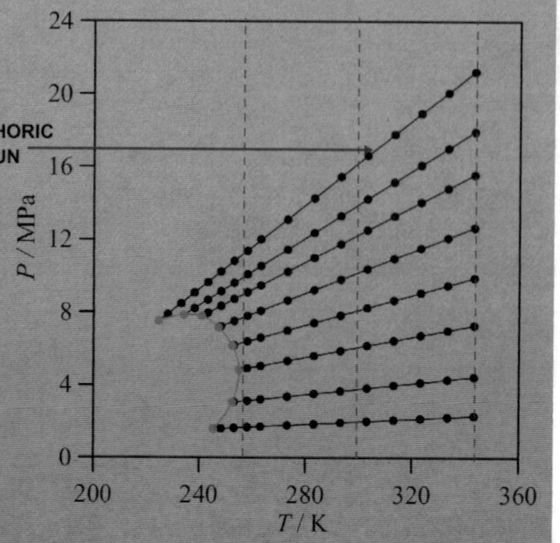

$$\frac{V(T,P)}{V_o(T_o,P_o)} = 1 + \gamma(P - P_o) + \beta(T - T_o)$$

Isochores → isochoric PVT apparatus
Isotherms → Magnetic Suspension Densimeter

ISOCHORIC EXPERIMENTAL DATA

Phase Envelope (SNG-3)		Phase Envelope (SNG-4)	
T / K	P / MPa	T / K	P / MPa
230.270	8.621	224.810	7.512
240.360	9.105	234.140	7.841
248.890	9.134	240.850	7.778
255.540	8.710	247.150	7.161
261.350	7.736	252.470	6.159
263.730	6.682	255.150	4.832
264.790	5.374	252.440	3.053
263.830	3.975	245.280	1.572
259.620	2.393		

Phase Envelope (SNG-5)		Phase Envelope (SNG-6)	
T / K	P / MPa	T / K	P / MPa
239.310	9.681	265.600	11.762
246.530	9.964	272.000	11.315
255.800	10.039	277.610	9.947
264.020	9.156	281.370	8.607
268.880	7.959	283.760	6.990
272.040	6.568	284.140	5.384
273.150	4.958	282.330	3.488
270.600	3.303	274.230	1.593

ISOCHORIC EXPERIMENTAL DATA

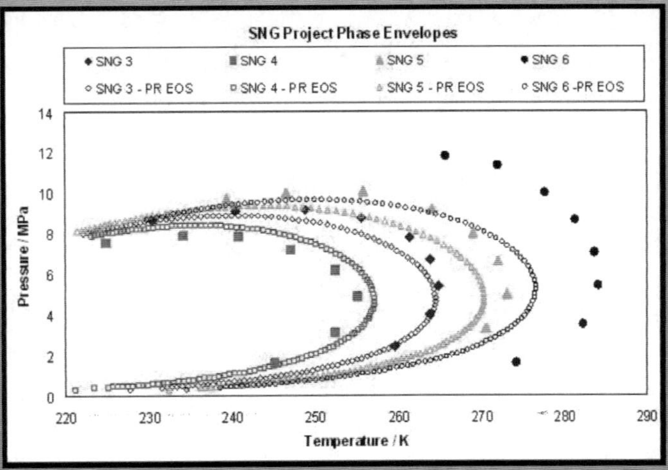

ISOCHORIC CALIBRATION

- 8 vapor pressures of CO_2 measured between 230 and 290 K.
  - maximum deviation of 0.055% w.r.t. Span et. al.[1]
  - maximum deviation of 0.040% w.r.t. correlation as implemented by Lemmon et al.
- 7 vapor pressures of C_3H_8 measured between 270 and 340 K.
  - maximum deviation of 0.07% w.r.t. Lemmon et.al.[2]

(1) Span, R. and Wagner, W., "A new equation of state for carbon dioxide covering the fluid region from the triple point temperature to 1100 K at pressures up to 800 MPa," *J. Phys. Chem. Ref. Data*, 25 (1996), 1509-1596.

(2) Lemmon, E.W., A.P. Peskin, M.O. McLinden, and D.G. Friend, NIST Standard Reference Database 12, National Institute of Standards and Technology, Standard Reference, Data Program, Gaithersburg, 2000.

MAGNETIC SUSPENSION DENSIMETER

Basis: **Archimedes' Principle:**

"A solid body (sinker) suspended in a fluid experiences a buoyant force exerted by the mass of a volume of fluid displaced by the solid body."

m_v = true mass of sinker (in vacuum)

m_a = apparent mass of sinker (in fluid

m_f = mass of fluid displaced by sinker

$$m_v > m_a$$
$$m_f = (m_v - m_a)$$

$$\rho_f = \frac{(m_v - m_a)}{v_S}$$

MAGNETIC SUSPENSION DENSIMETER

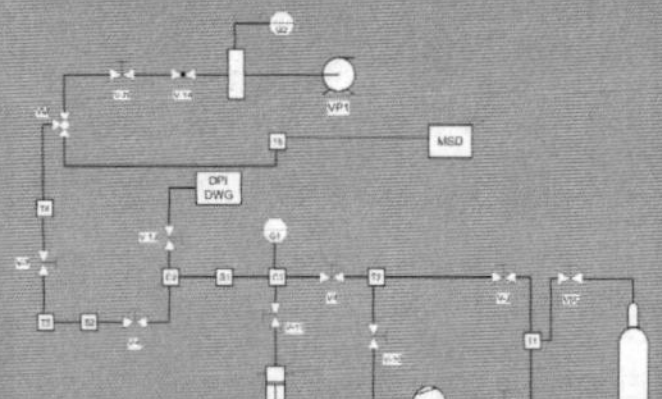

Pressurizing Manifold

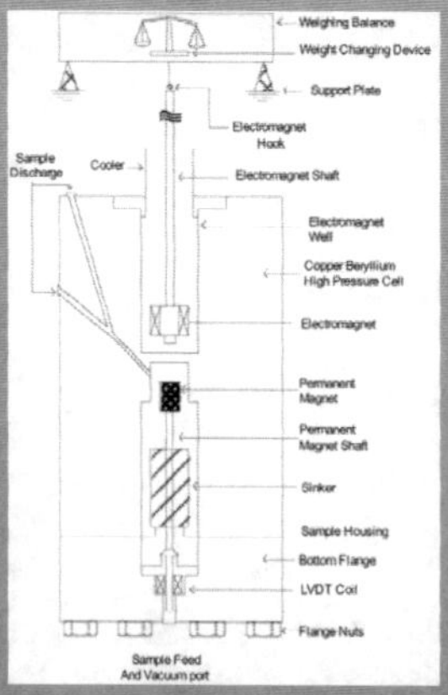

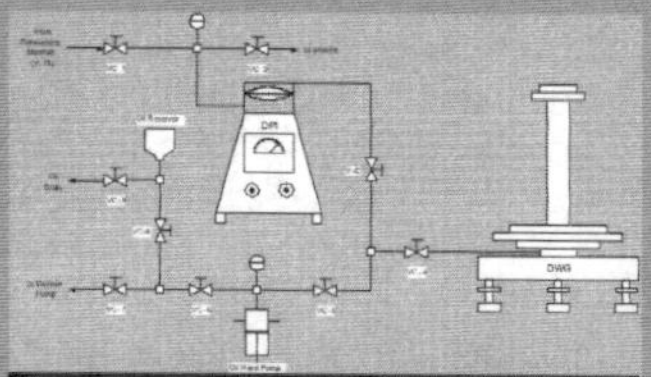

Pressure Transducer Calibration

Compressor Unit

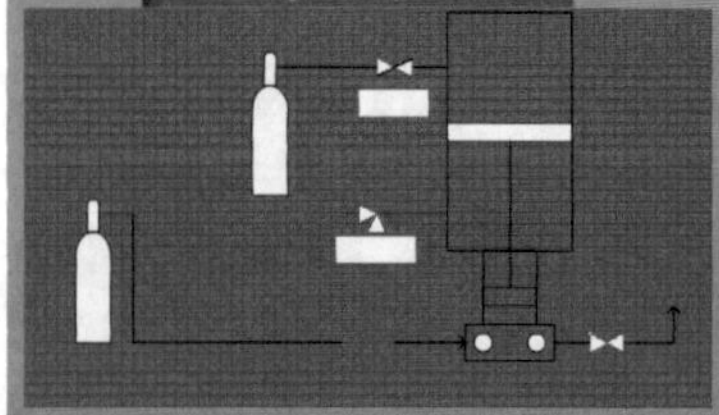

Pressure Monitoring

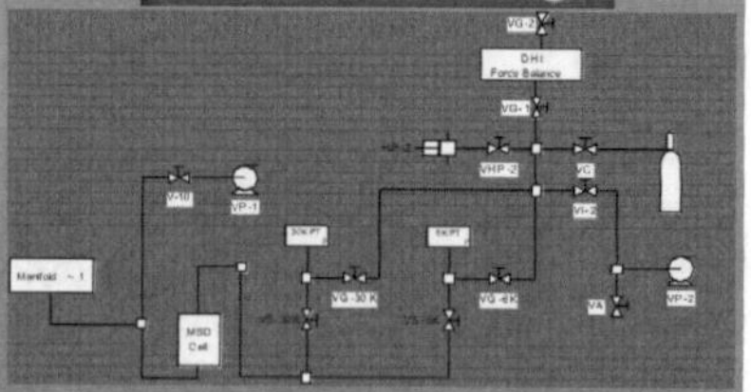

MAGNETIC SUSPENSION DENSIMETER

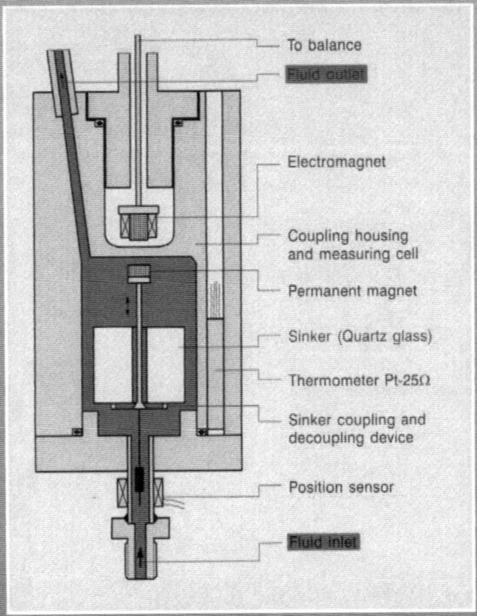

Weighing Balance

- Range: 0 – 200 g
- Accuracy: 0.003% reading
- Resolution: 0.01 mg

Pressure Transducer

- Range: 0 – 207 MPa (30000 psia)
- Accuracy: 0.02% full scale

Densimeter

Range: 0 – 2000 kg/m^3

- up to 2000 bar / -80 to 100°C
- up to 1300 bar / -80 to 250°C

Accuracy: 0.03 % or 0.005 kg/m^3

PRT

- Range: -80 to 250°C
- Accuracy: 5mK

CALIBRATION - PURES

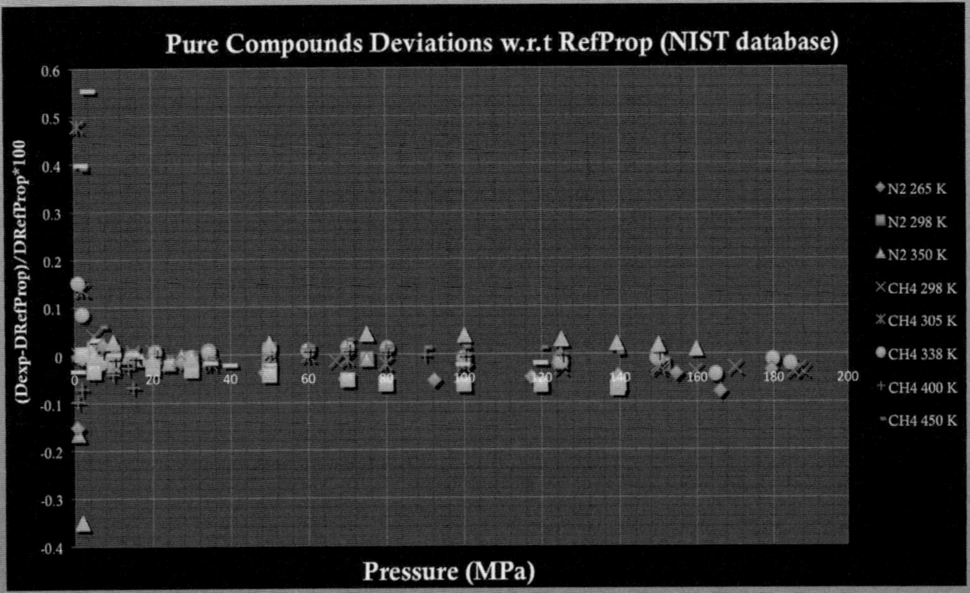

SNG3 MIXTURE

Component	Mol %
Methane	89.982
Ethane	3.009
Propane	1.506
Iso-Butane	0.752
N-Butane	0.753
Iso-Pentane	0.300
N-Pentane	0.300
Nitrogen	1.697
Carbon Dioxide	1.701

3 isotherms at 250K, 350K and 450K

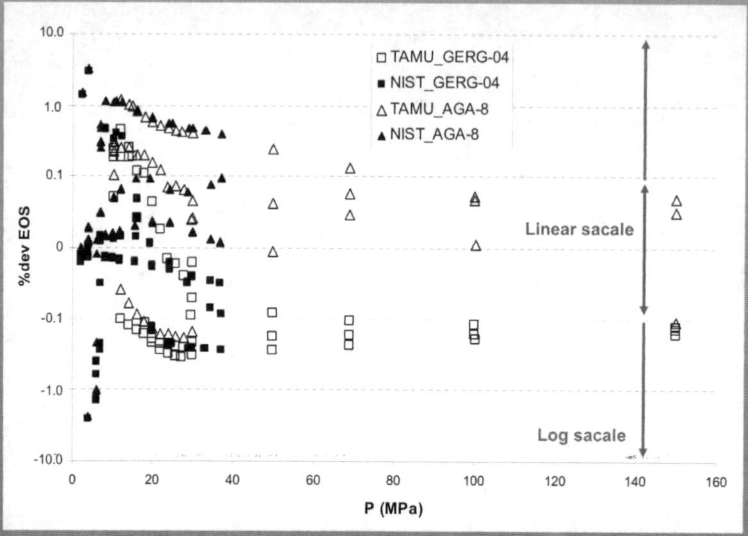

James C. Holste; Kenneth R. Hall; Gustavo A. Iglesias-Silva, AIChE Journal 1996, 42 (1), 296-297.

SNG4 MIXTURE

Component	Mol %
Methane	89.900
Ethane	3.150
Propane	1.583
Iso-Butane	0.781
N-Butane	0.790
Iso-Pentane	0.150
N-Pentane	0.150
Nitrogen	1.699
Carbon Dioxide	1.707

3 isotherms at 250K, 350K and 450K

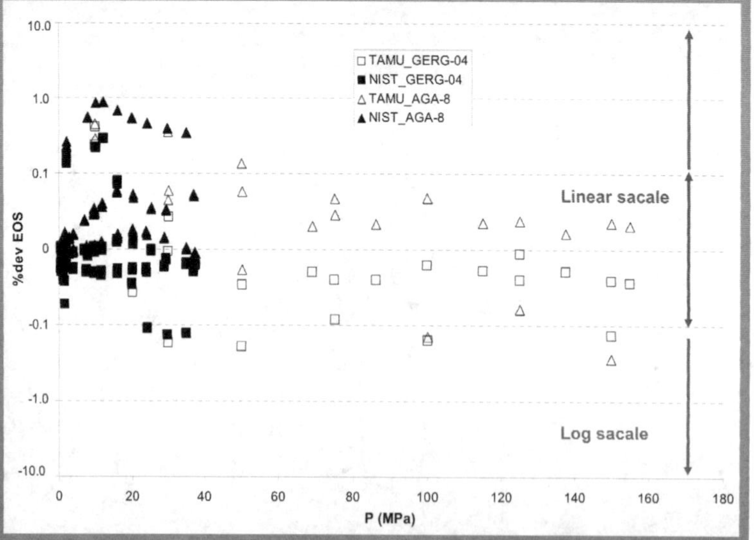

James C. Holste; Kenneth R. Hall; Gustavo A. Iglesias-Silva, AIChE Journal 1996, 42 (1), 296-297.

SNG5 MIXTURE

Component	Mol %
Methane	89.975
Ethane	2.855
Propane	1.427
Iso-Butane	0.709
N-Butane	0.722
Iso-Pentane	0.450
N-Pentane	0.450
Nitrogen	1.713
Carbon Dioxide	1.699

3 isotherms at 250K, 350K and 450K

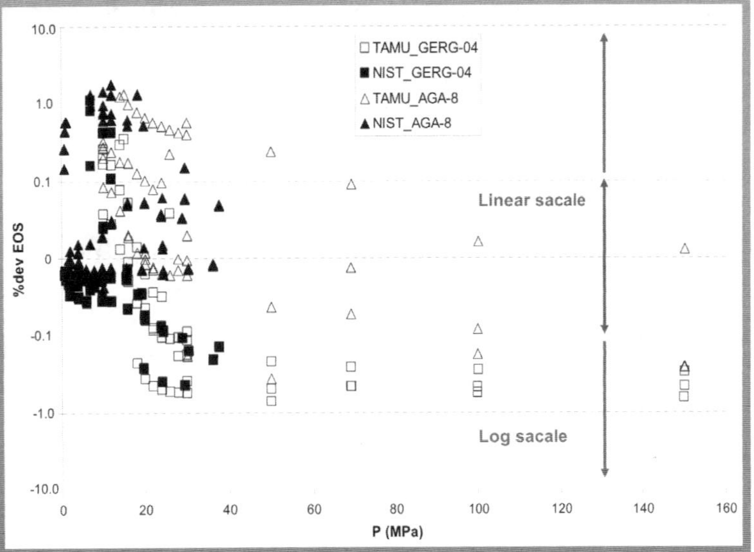

James C. Holste; Kenneth R. Hall; Gustavo A. Iglesias-Silva, AIChE Journal 1996, 42 (1), 296-297.

SNG6 MIXTURE

Component	Mol %
Methane	90.001
Ethane	4.565
Propane	2.243
Iso-Butane	1.140
N-Butane	1.151
Iso-Pentane	0.450
N-Pentane	0.450

3 isotherms at 250K, 350K and 450K

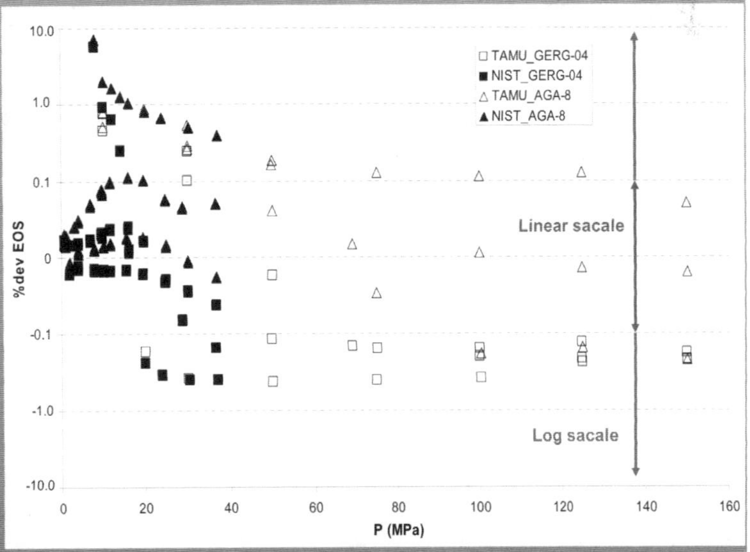

James C. Holste; Kenneth R. Hall; Gustavo A. Iglesias-Silva, AIChE Journal 1996, 42 (1), 296-297.

CONCLUSIONS

- **TAMU density data are consistent with NIST data (double sinker MSD) in the low pressure range.**
- **MSD density data agree with standard equations of state within their uncertainty limits, except for a few points in the low pressure range.**
- **Phase loop predictions from the widely used PR-EoS show large deviations w.r.t. experimental data near both cricondentherm and cricondenbar.**
- **As the long chain HC concentration increases PR-EoS tends to under-predict CT and CB regions.**

Jens Ahlers*, Jürgen Rarey*, Jürgen Gmehling* Todd Willman**
Chair of Industrial Chemistry University of Oldenburg, Germany * DDBST GmbH * EPCON international**

CARL VON OSSIETZKY universität OLDENBURG

DDBST DORTMUND DATA BANK SOFTWARE & SEPARATION TECHNOLOGY

EPCON International

Improving your separation process simulation with accurate binary parameters

AIChE Regional

Process Technology Conference

Galveston, TX October 2009

Problems of fitting g^E model parameters only to isobaric VLE data

H. C. Van Ness (Rossini Lecture 1994)

Pure & Appl. Chem. 67, 859 (1995)

In the early unsophisticated days of chemical engineering VLE data were taken at constant pressure for direct application in the design of distillation columns, which were treated as though they operated at uniform pressure. There is no longer any excuse for taking isobaric data, but regrettably the practice persists, and we have examined two such data sets. Rigorous thermodynamic treatment of isobaric data present problems that do not arise with isothermal data. Their origin is the need to take into account not only the composition dependence of the excess Gibbs energy but also its temperature dependence.

Results for the system ethanol (1) – n-decane (2) using Wilson parameters fitted to isobaric VLE only and recommended values

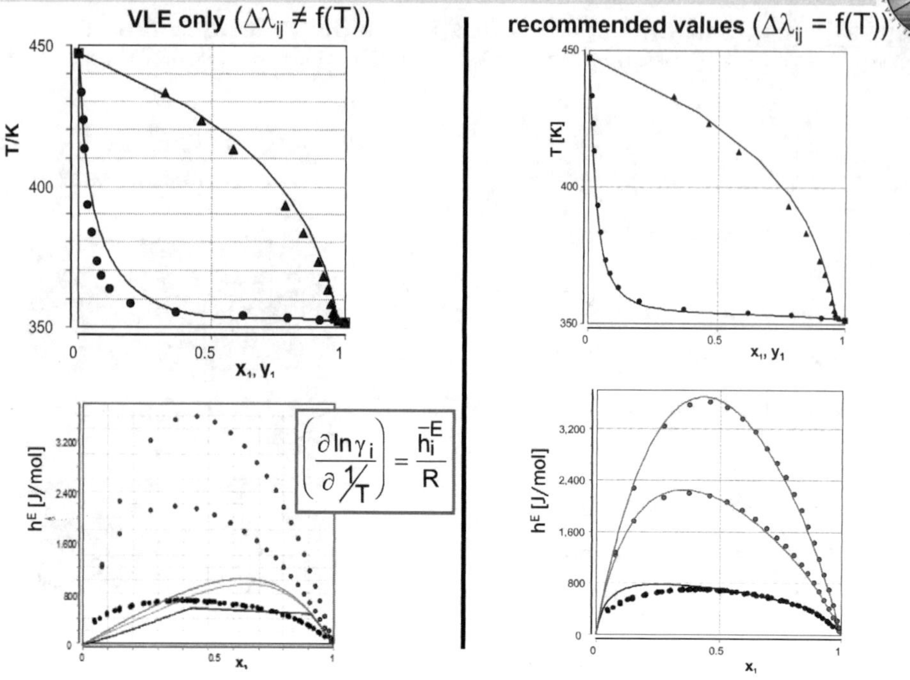

Results for the system ethanol (1) – n-decane (2) using Wilson parameters fitted to isobaric VLE only and recommended values

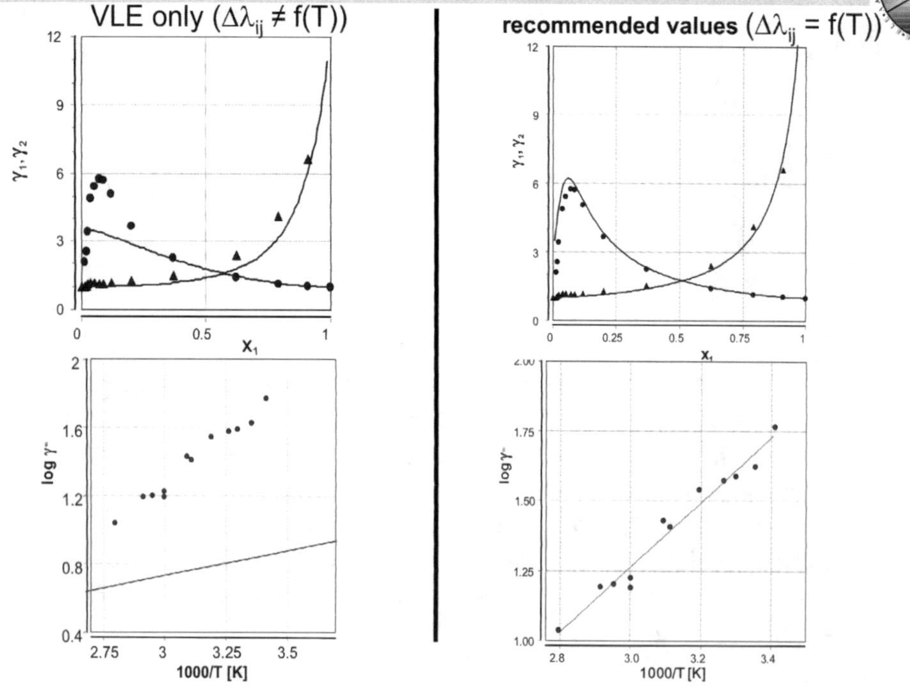

VLE at 101.3 kPa of Acetone (1) + Water (2) using NRTL parameters taken from process simulator

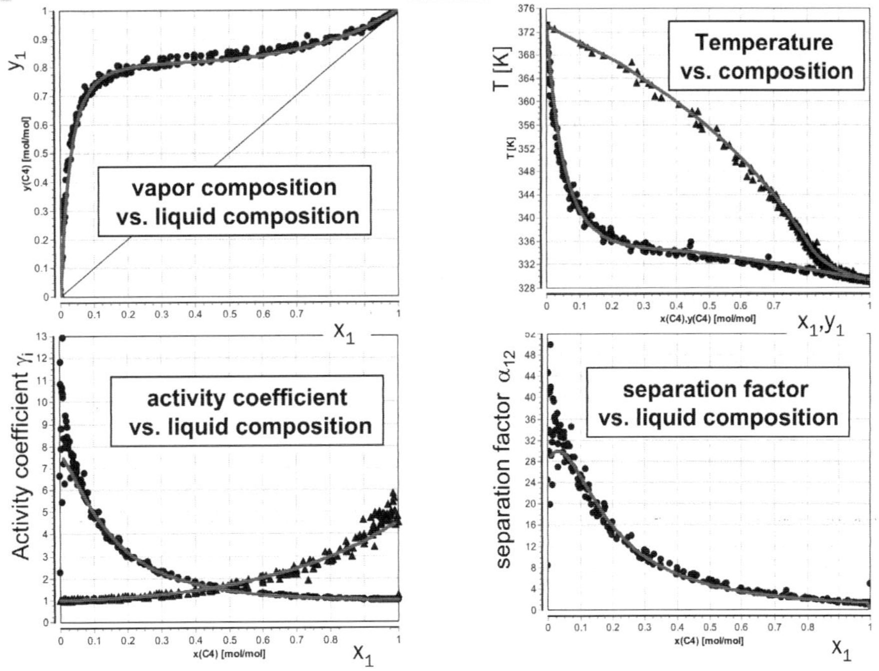

VLE of Acetone (1) + Water (2) at 101.3 kPa using simulator parameters

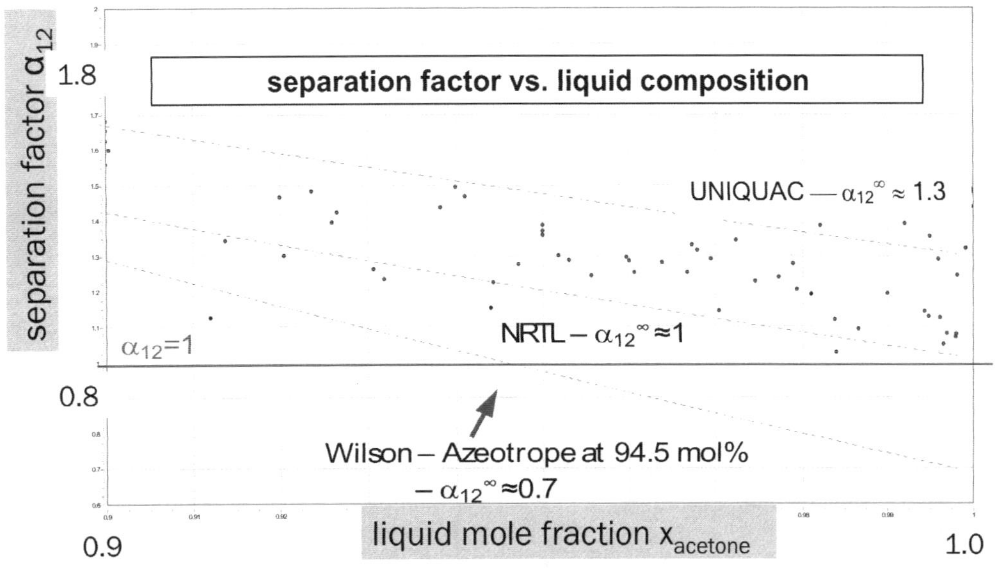

Activity coefficients at infinite dilution of Acetone + Water using NRTL

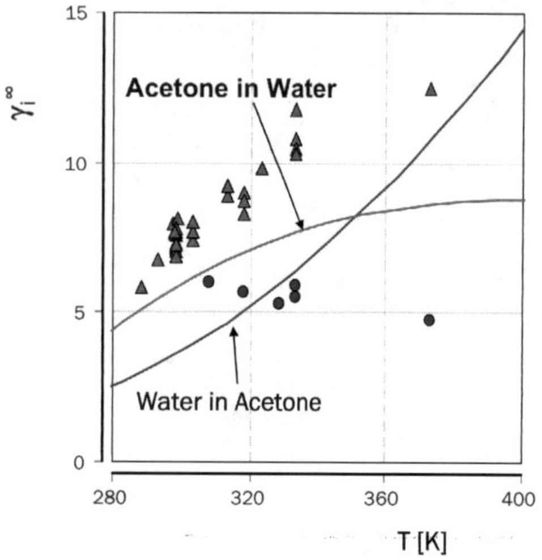

NRTL using **simulator** parameters

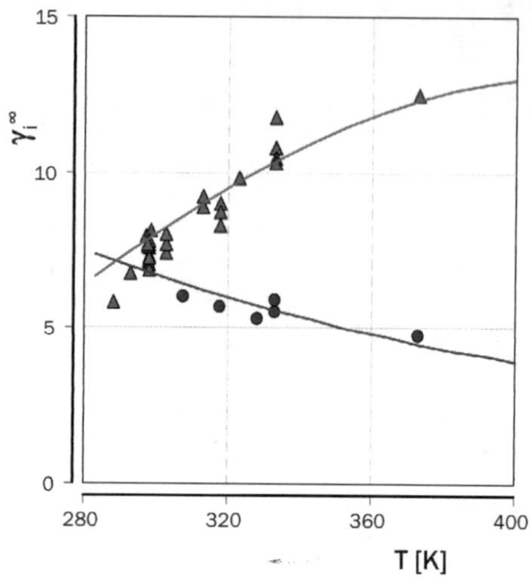

NRTL parameters from simultaneous regression to VLE, h^E, γ^∞ and azeotropic data

Excess enthalpies of Acetone + Water using NRTL

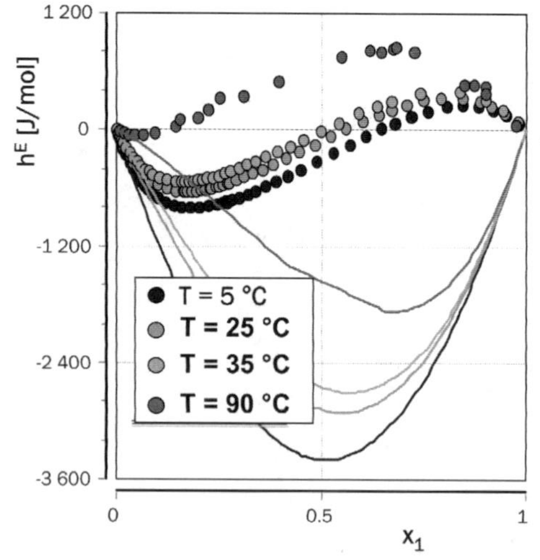

NRTL using **simulator** parameters

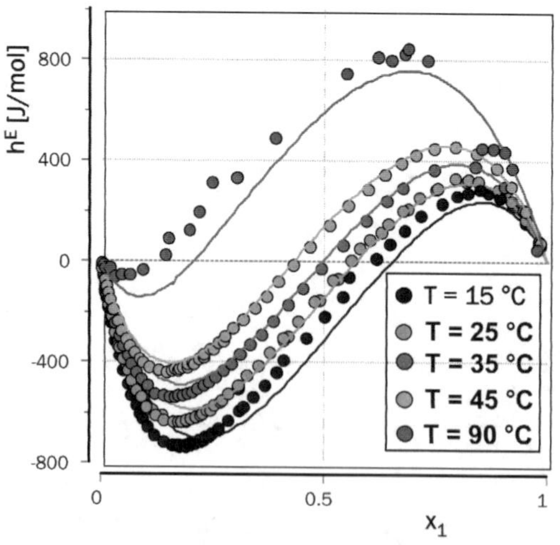

NRTL parameters from simultaneous regression to VLE, h^E, γ^∞ and azeotropic data

Azeotropic data at infinite dilution of Acetone + Water using NRTL

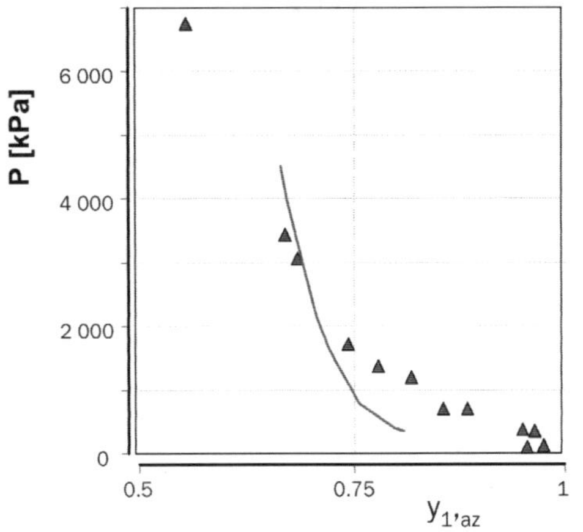

NRTL using **simulator** parameters

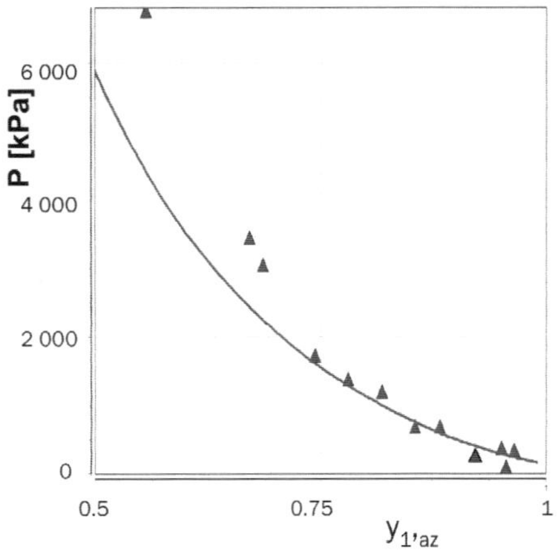

NRTL parameters from simultaneous regression to VLE, h^E, γ^∞ and azeotropic data

VLE at 101.3 kPa of Acetone (1) + Water (2) using NRTL comparison simulator - revised regression

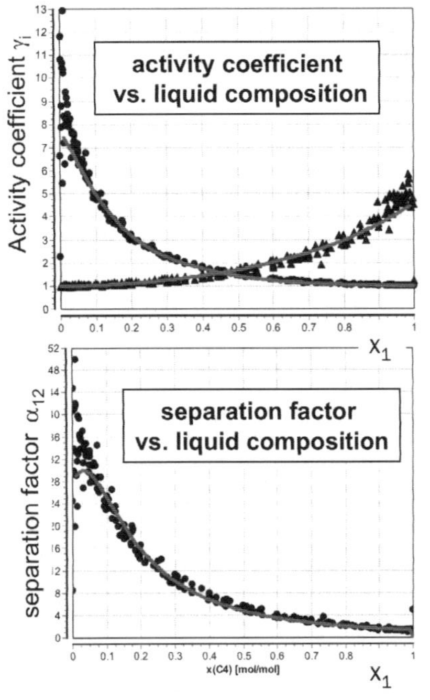

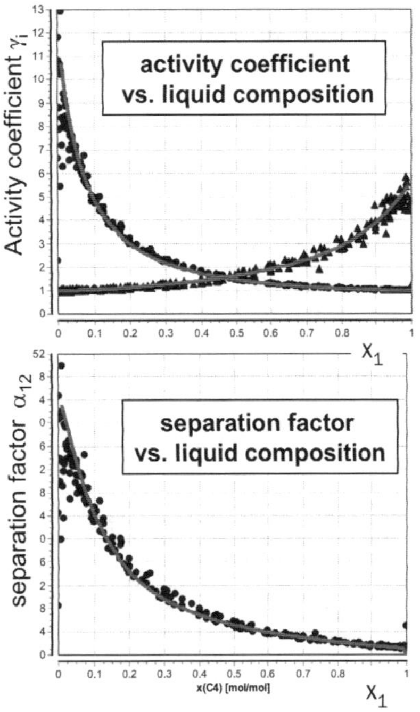

DDB: An Ideal Tool for the Critical Examination of Model Parameters Prior to Process Simulation (Acetone (1) + Cyclohexane (2) (Wilson))

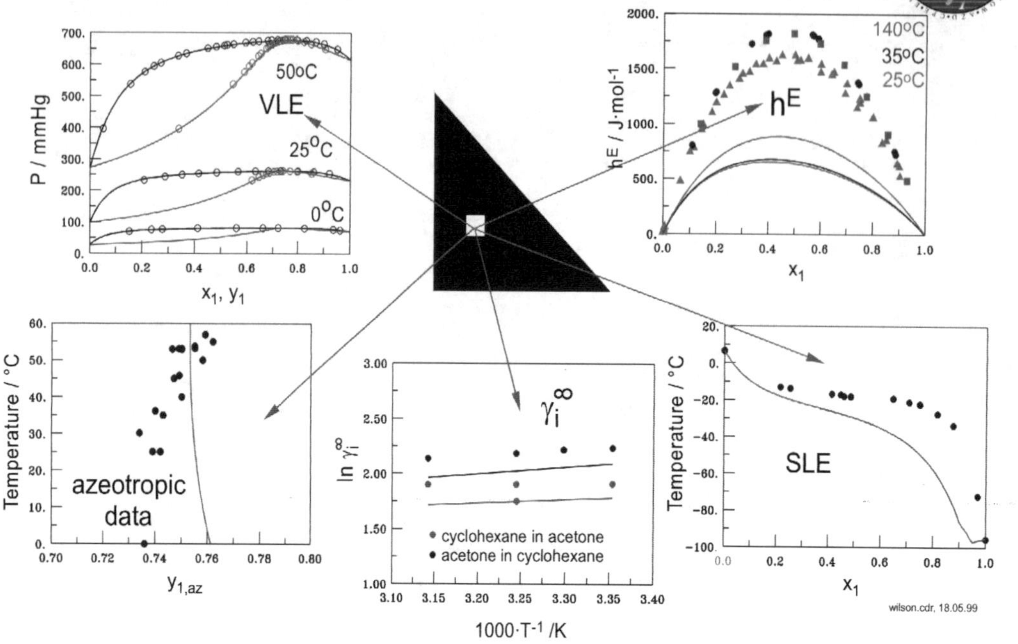

The GPA Data Bank – Joint DDBST & EPCON Project for GPA

DDBST DORTMUND DATA BANK SOFTWARE & SEPARATION TECHNOLOGY

EPCON International

GPA

- The new GPA Data Bank, Fifth Edition, provides 100% of all published GPA experimental thermophysical property data in electronic format including searchable text copies of all GPA Research Reports and Technical Publications.
- A joint project undertaken by DDBST GmbH (Oldenburg, Germany) and EPCON International (Houston, TX) with funding by GPA over a 3 year period.
- The result is a comprehensive compilation of all GPA experimental data including VLE data for low boiling hydrocarbons available in standardized Dortmund Data Bank format that is easily accessed by many process simulation software products.
- Included is regression software allows for generation of binary parameters for direct use in process simulation software including NRTL, Peng Robinson, and SRK methods along with comprehensive data searching and graphing capabilities.

GPA Data Bank – Main Applications

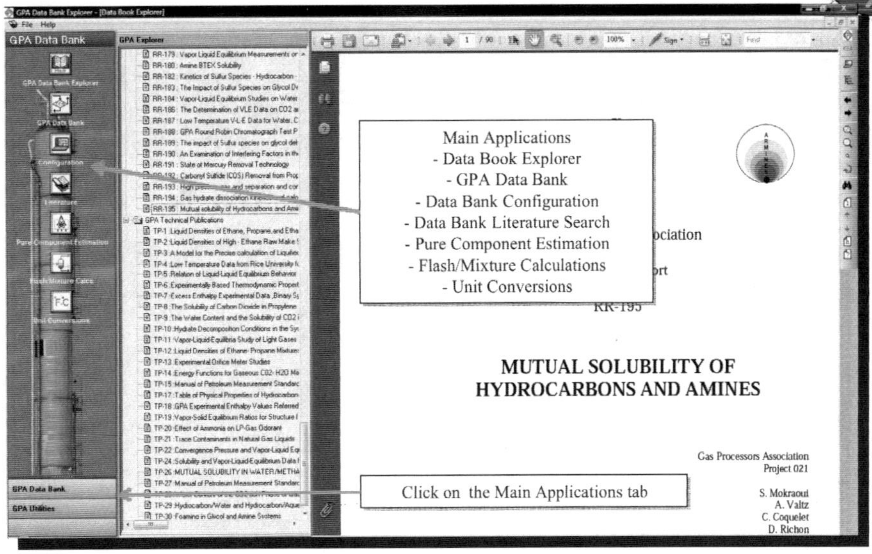

The navigation bar on the left side of the screen is used to access programs in the GPA Data Bank. The Main Applications are the primary software tools developed by DDBST and EPCON to view, graph, and fit data for determination of phase equilibrium for any gas composition. Temperature and pressure. The binary parameters generated can be used to improve your process simulation results.

GPA Data Bank – Utility Programs

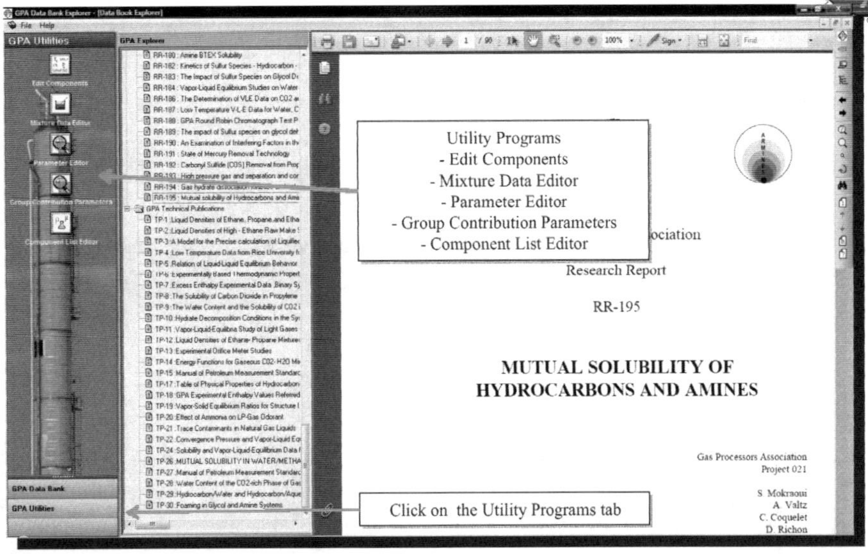

The Utility Programs in the GPA Data Bank provide the ability to edit component and mixture data. The Parameter Editor is used to store the binary parameters generated from the GPA Data Bank for direct use in process simulation software.

GPA Data Bank – Data Bank Explorer

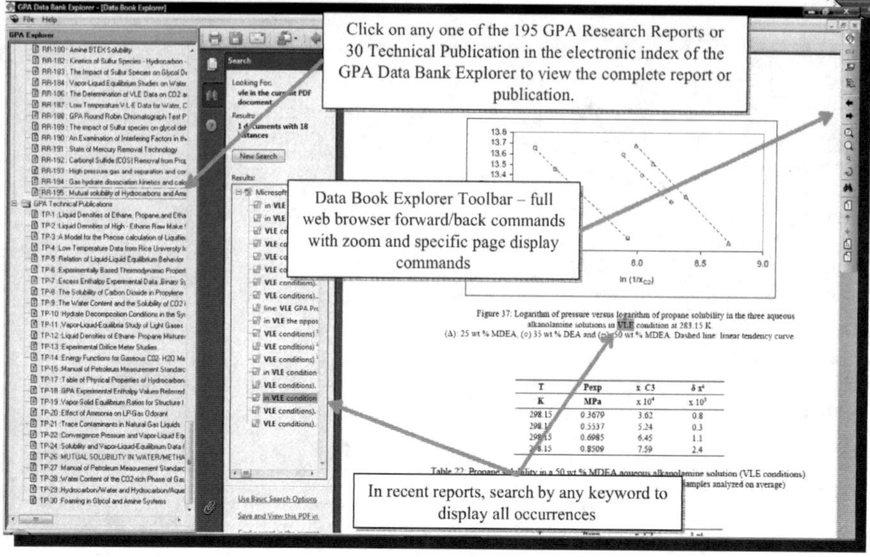

The GPA Data Bank provides all of the GPA Research Reports and Technical Publications in electronic, searchable format using the Data Bank Explorer interface. This provides quick access to over 30 years of GPA research work with all experimental data converted into electronic data format - providing full value from your company's GPA research investments.

GPA Data Bank – Component Selection

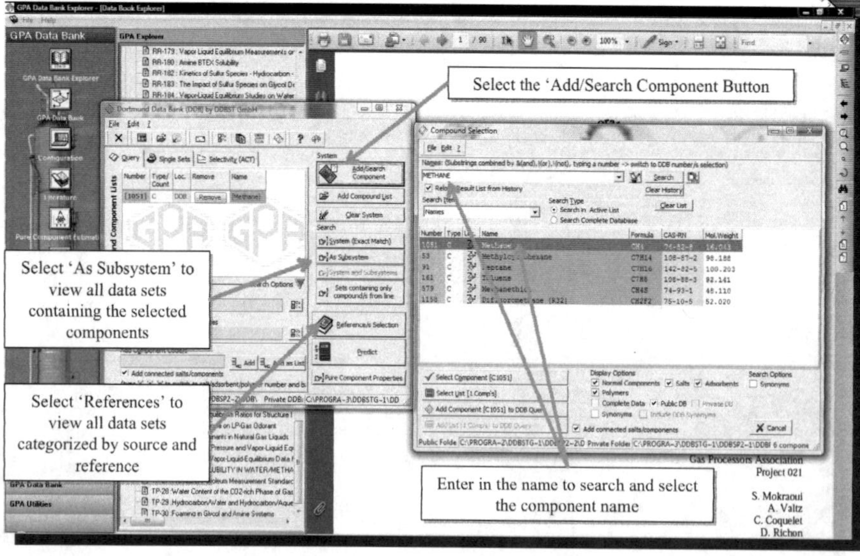

All GPA components can be searched in a number of ways for retrieval of pure and mixture data. Selection of 'System (Exact Match)' will display all systems where only the selected components appear while selection of 'As Subsystem' will display all systems where the selected components and all others appear. VLE data for these systems is then automatically selected and graphed for review.

GPA Data Bank – Automated Displays and Graphs

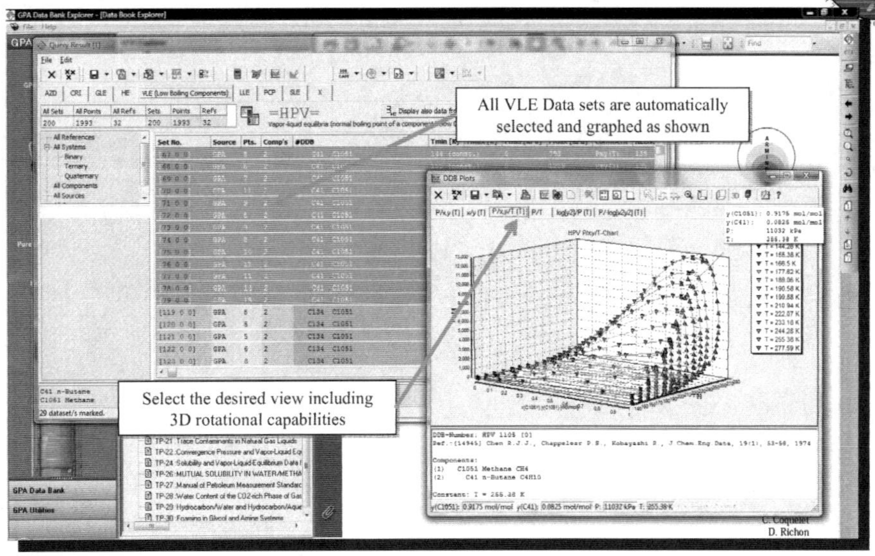

The Query Results window shows all data sets for each data type as shown on the property tabs (the default selection is VLE data). The graphing of VLE data is done automatically and other data types can also be graphed. As the pointer is moved over a data point on the graph, the numeric data is displayed in a floating window as shown. Data can be exported in a several file formats.

GPA Data Bank – Fitting Binary Parameters

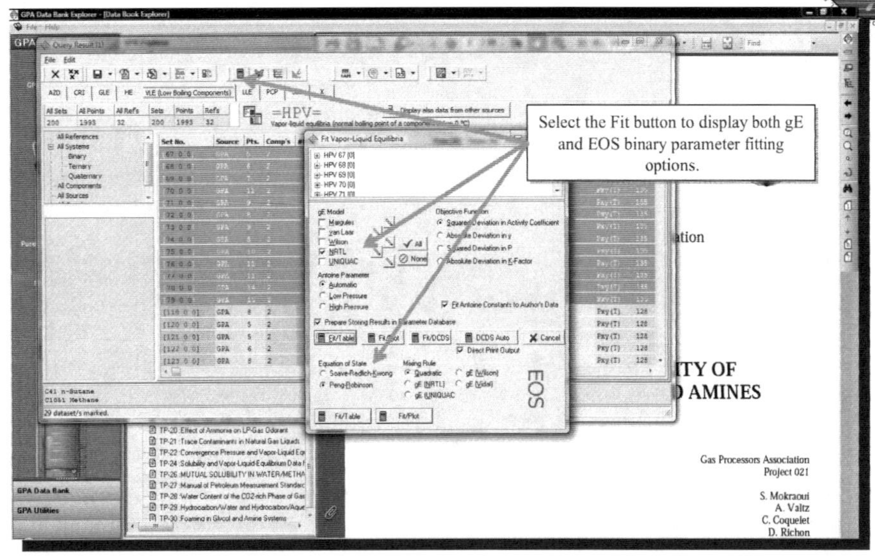

Automated fitting of binary parameters for either gE and EOS models is provided for selected data sets. View the fit graphically or numerically and save parameters to the Parameter database for use in process simulation.

GPA Data Bank – VLE Prediction using UNIFAC

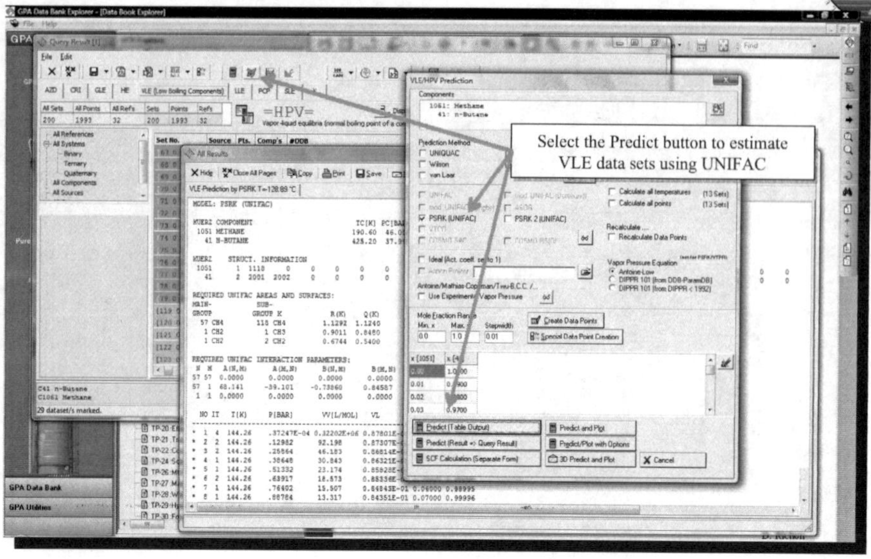

Prediction of VLE data sets using PSRK and UNIFAC can also be provided by the GPA Data Bank. The VLE data sets can then be used for prediction of binary parameters.

GPA Data Bank – Access to All Experimental Data

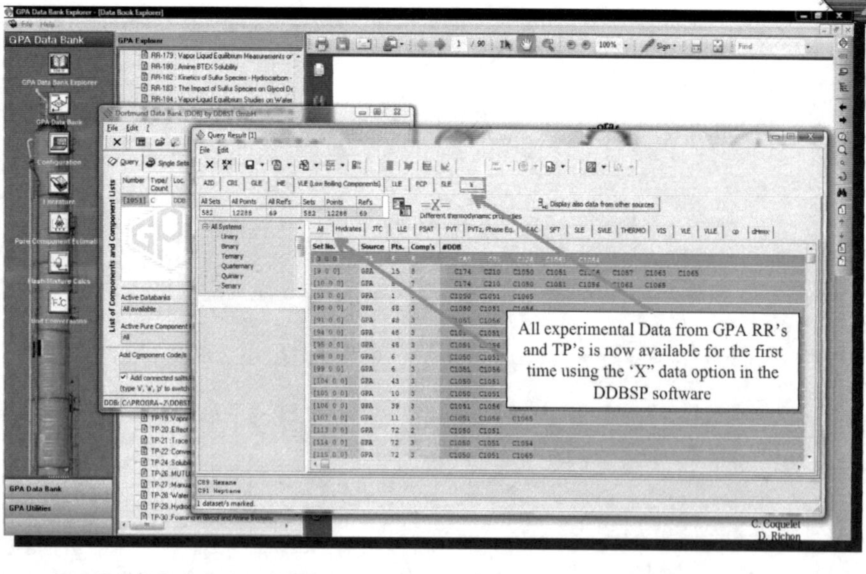

The Dortmund Data Bank Software Package (DDBSP) basic software applications are included in the GPA Data Bank. The DDB data types were extended with the 'X' data type so that all GPA experimental data could be captured into the GPA Data Bank. Selection of the 'X' data type provides a secondary set of property tabs that may be chosen to view the data sets available.

GPA Data Bank – Access to References and Data

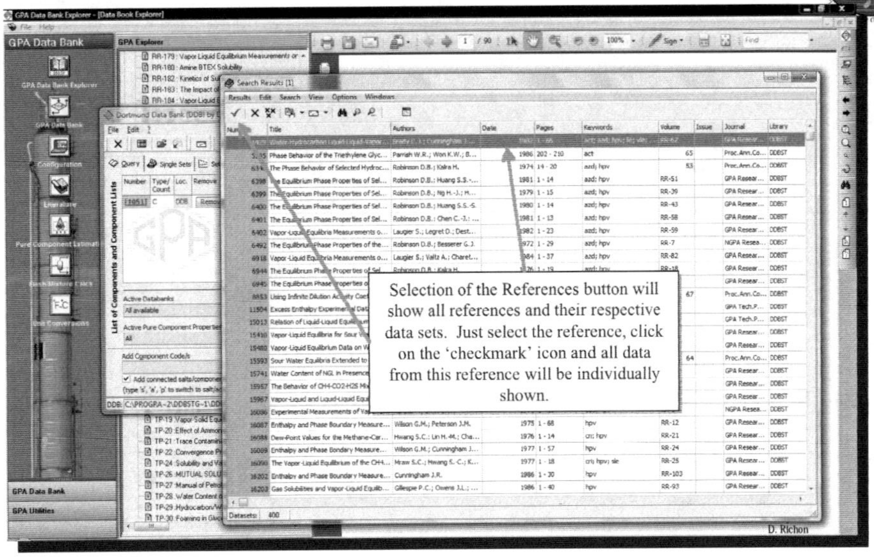

Selection of the References button in the GPA Data Bank main window will provide a list of all references with authors, dates, volumes, etc. Selection of one of these references and clicking on the 'checkmark' icon will display all the data sets in that reference for further graphing, fitting, etc. in the Query Results window.

GPA Data Bank – ARTIST Property Estimation

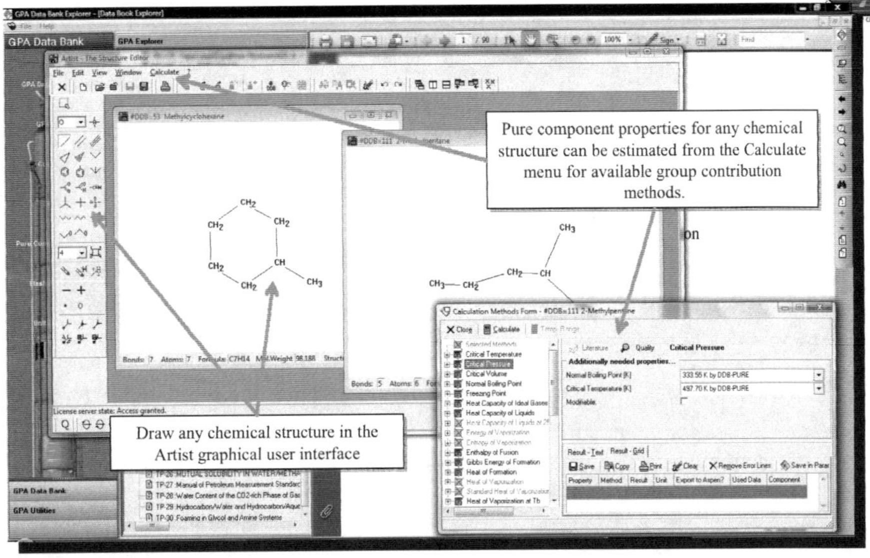

The Artist graphical user interface allows for drawing of any chemical structure and estimation of pure component properties. Group contribution methods are provided for basic estimation capabilities and extended methods are available separately.

GPA Data Bank – Flash Mixture Calculations

Enter Feed Conditions and Flash Conditions

Select to run flash calculations

Enter pure and pseudo components

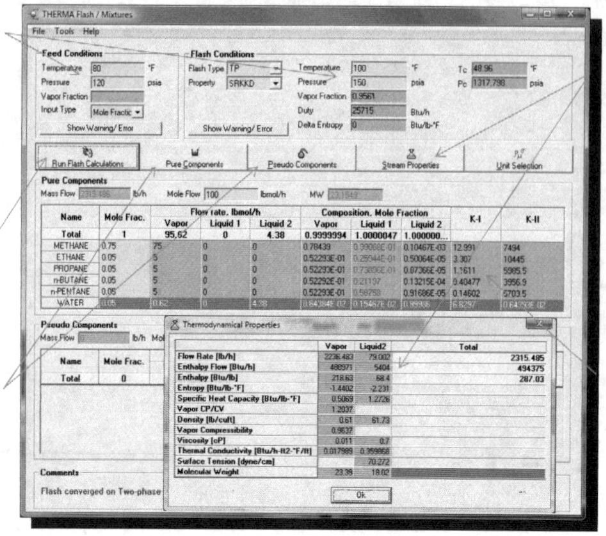

View Stream Properties when calculation is completed

Phase flows and compositions are shown for 3-phases

The Flash/Mixture Calculations program provides phase equilibrium and physical properties for each phase. The prediction of the second liquid (aqueous) phase is highly accurate based on its development and use in the API Technical Data Book software available separately. Use of binary parameters developed in the GPA Data Bank allow for highly accurate phase equilibrium (flash) calculations.

DDB Simulation Software Thermo Engine Options

Calculation using Aspen Property Toolkit to verify physical property issues prior to process simulation.

2008 VLE, HPV

2009 h^E, c_P^E, v^E

•Aspen Plus completed

•Pro/II-Interface mostly finished.

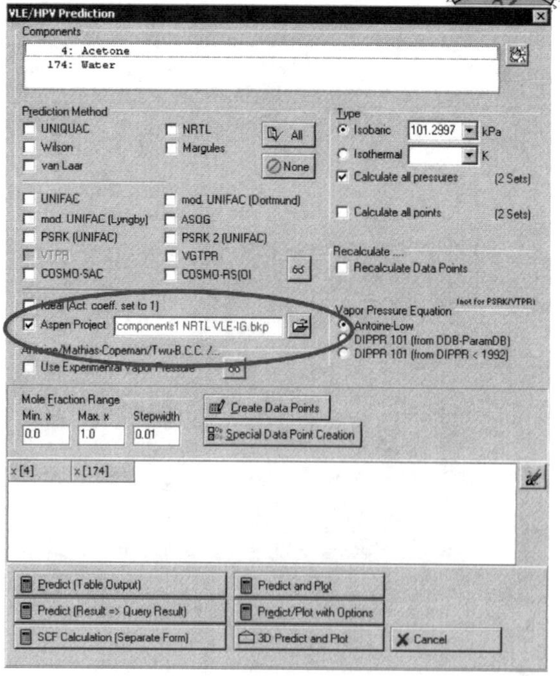

DDBSP & GPA Data bundled in SYSTEM 7 Process Explorer®

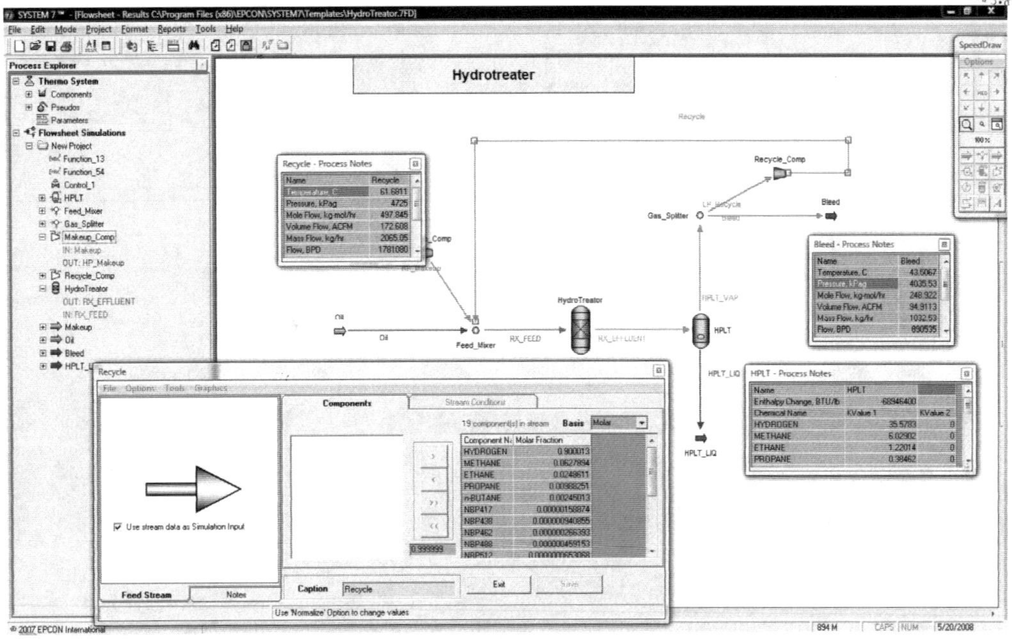

(For more information, DDBST is represented in the EPCON International booth)

Jens Ahlers, Jürgen Rarey, Jürgen Gmehling

Chair of Industrial Chemistry
University of Oldenburg, Germany
DDBST GmbH

CARL VON OSSIETZKY universität OLDENBURG

Present status and potential of group contribution methods for process development

AIChE Regional
Process Technology Conference
Galveston, TX October 2009

Prof. Gmehling Rossini Lecture Award 2008

INTERNATIONAL ASSOCIATION of CHEMICAL THERMODYNAMICS

In recognition of outstanding contributions to chemical thermodynamics

Professor Jürgen Gmehling

was elected as

ROSSINI LECTURER

in 2008

John H. Dymond, Secretary — Jean-Pierre E. Grolier, Chairman

Overview on the Requirement and the Available VLE Data for Binary Systems (2008, 1973)

Assuming that 1000 compounds are of technical interest VLE data for appr. 500 000 binary systems are required

Number of VLE Data Sets stored in the Dortmund Data Bank:

55 450 isothermal / isobaric VLE data sets

(but 240 data sets for ethanol-water, 150 data sets for acetic acid-water, 175 data sets for methanol-water, ..

⟶ VLE data for 10 520 binary systems: 2.10 %

Situation 1973: 17600 isothermal / isobaric data sets

⟶ VLE data for 3 803 binary systems: 0.76 %

06.02.0

Group Contribution Method

gl_e.cdr

ation Symposium Brighton 1969

; (solution of groups concept)

$$\ln\gamma_i = \ln\gamma_i^C + \ln\gamma_i^R$$

Ethanol:

n-Hexan

Gre
Num
Req

Prof. Dr. Ulfert Onken

(University of Dortmund)

Experimental and Predicted Vapor-Liquid Equilibria Using UNIFAC

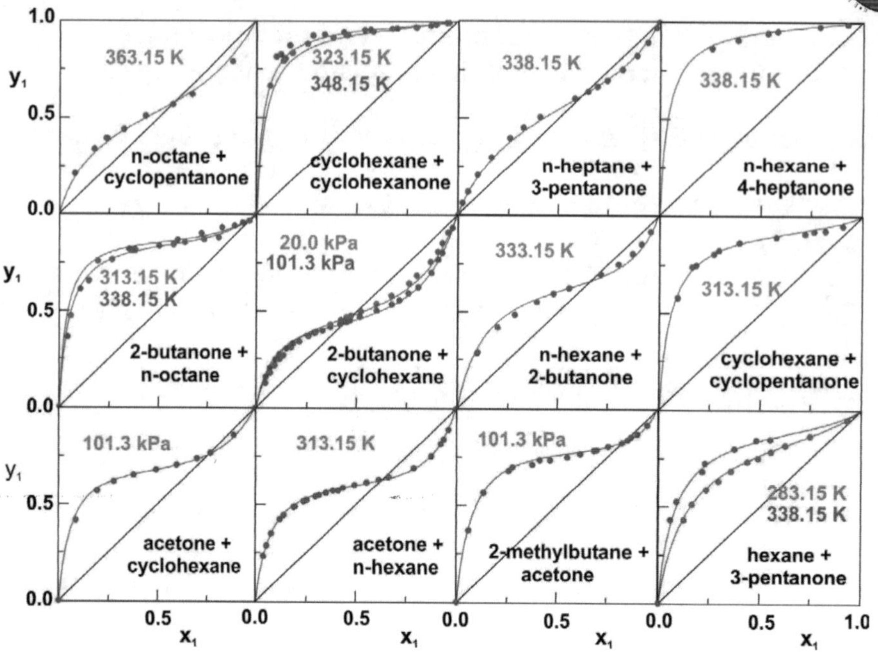

12.01.00

From **UNIFAC** to mod. UNIFAC $F = \sum VLE\ (+ \sum LLE + \sum \gamma^{\infty})$

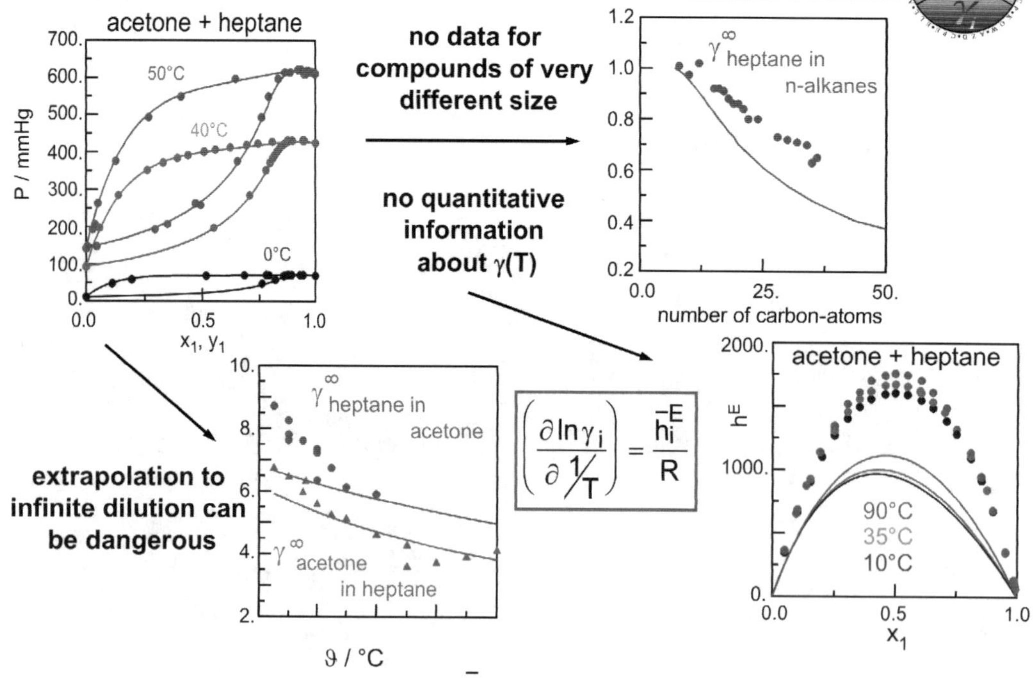

no data for compounds of very different size

no quantitative information about γ(T)

extrapolation to infinite dilution can be dangerous

$$\left(\frac{\partial \ln \gamma_i}{\partial \frac{1}{T}}\right) = \frac{\bar{h}_i^E}{R}$$

12.01.00

From UNIFAC to **mod. UNIFAC** (U. Weidlich, J. Gmehling 1987)

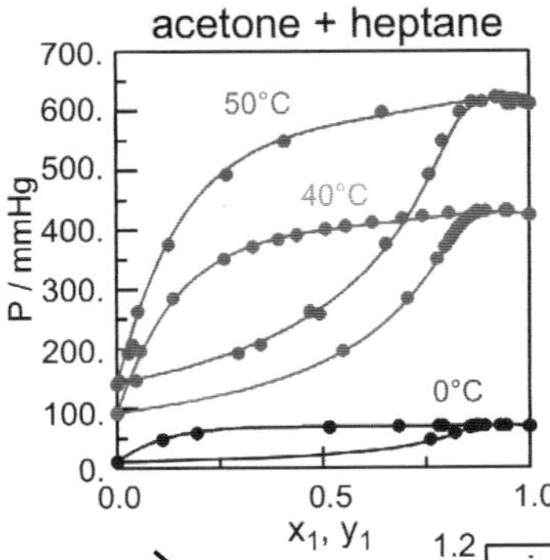

mixture information for the dilute region is used for fitting

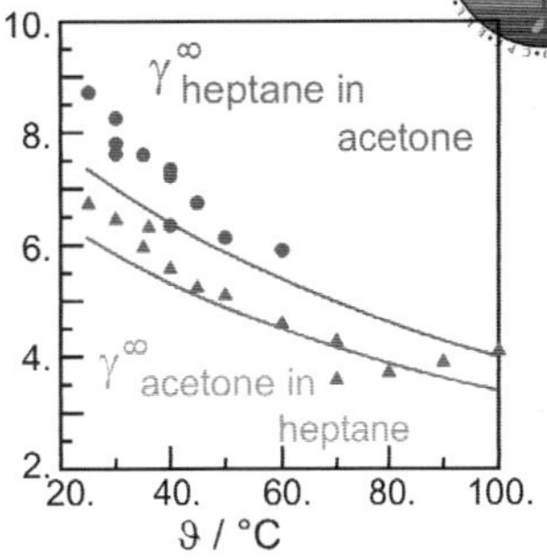

quantitative information about γ(T) from excess enthalpies

improved results for asymmetric mixtures using a modified combinatorial part

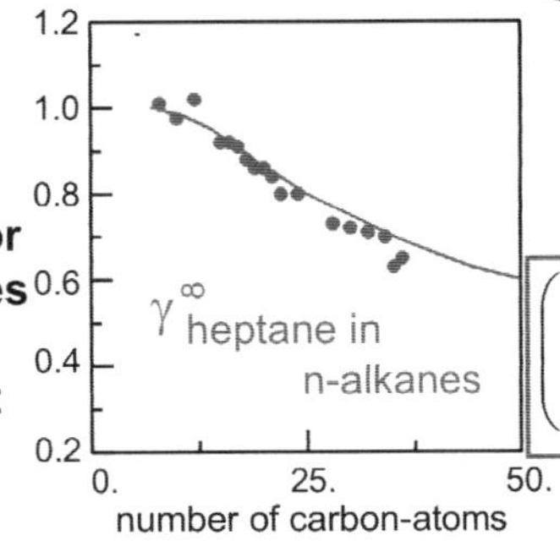

$$\left(\frac{\partial \ln \gamma_i}{\partial \frac{1}{T}}\right) = \frac{\bar{h}_i^E}{R}$$

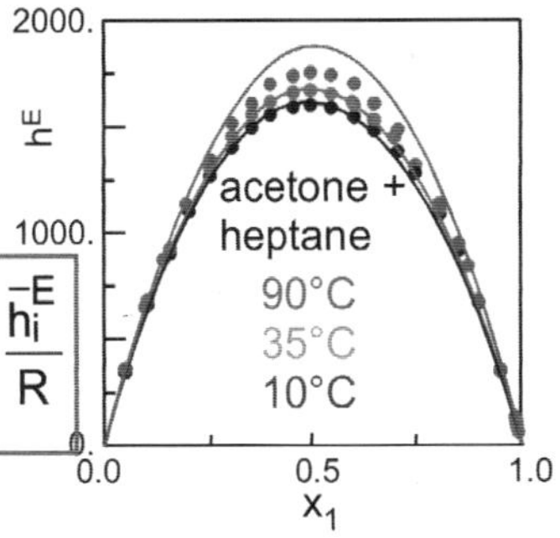

10.02.03

Data Base (Objective Function) Used for Fitting Modified UNIFAC (Dortmund) Parameters

Objective Function:

$$F = w_{VLE}\sum \Delta VLE + w_{AZD}\sum \Delta AZD + w_{h^E}\sum \Delta h^E + w_{c_p^E}\sum \Delta c_p^E + w_{\gamma^\infty}\sum \Delta\gamma^\infty + w_{LLE}\sum \Delta LLE + w_{SLE}\sum \Delta SLE$$

mixture data type		delivers the required information
VLE, azeotropic data		$\gamma = f(x)$
h^E (c_P^E)	Gibbs-Helmholtz equation $\frac{\partial \ln \gamma_i}{\partial (1/T)} = \frac{\bar{h}_i^E}{R}$	$\gamma = f(T)$ supporting data at high temperature
γ^∞		the only reliable information in the dilute range and about asymmetric systems
SLE (eutectic systems)	$\ln x_i^L \gamma_i^L = -\frac{\Delta h_{m,T_{m,i}}}{RT}\left(1 - \frac{T}{T_{m,i}}\right)$	supporting data at low temperature (T < 273 K)
LLE		often the only information for strong real mixtures

Experimental and Predicted Mod. UNIFAC (Do) Results for Different Alkane - Ketone - Systems

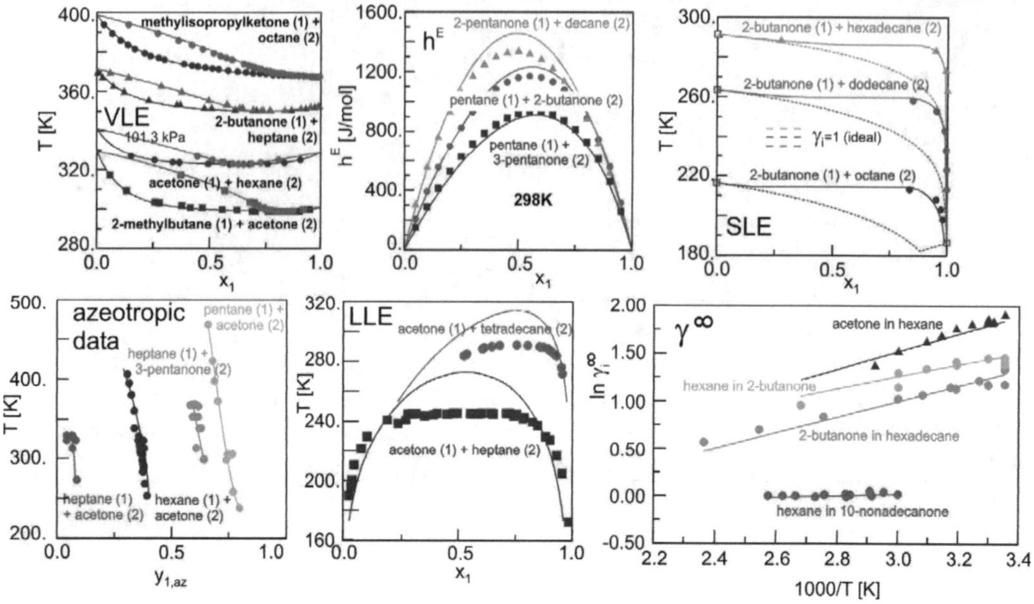

DDB 2009 – The Growth

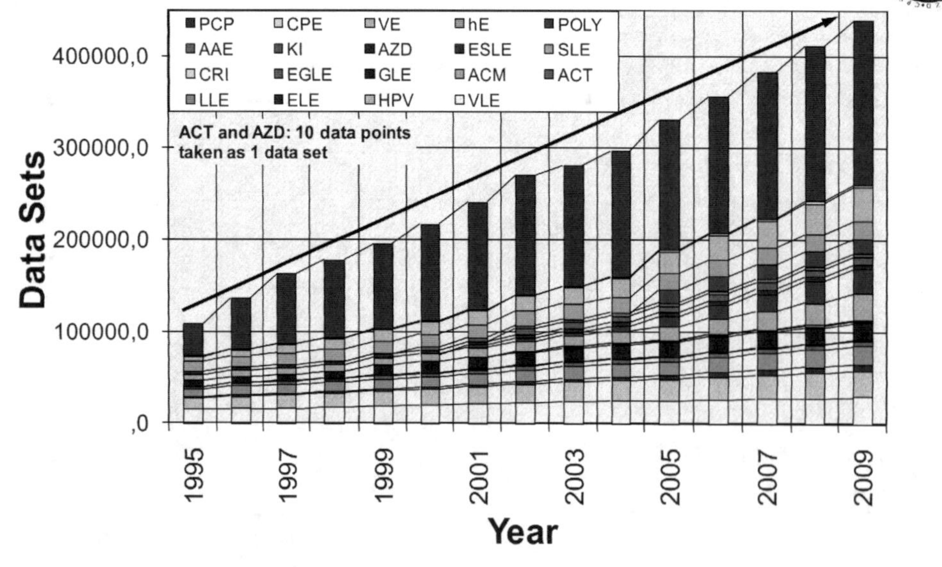

≈140 data sets per working day (yearly increase last 2 years ≈ 6.5 %)

Status of the Dortmund Data Bank* (April 2009)

61000 References, 1800 Journals, 28600 Compounds incl. Salts, Adsorbents and Polymers

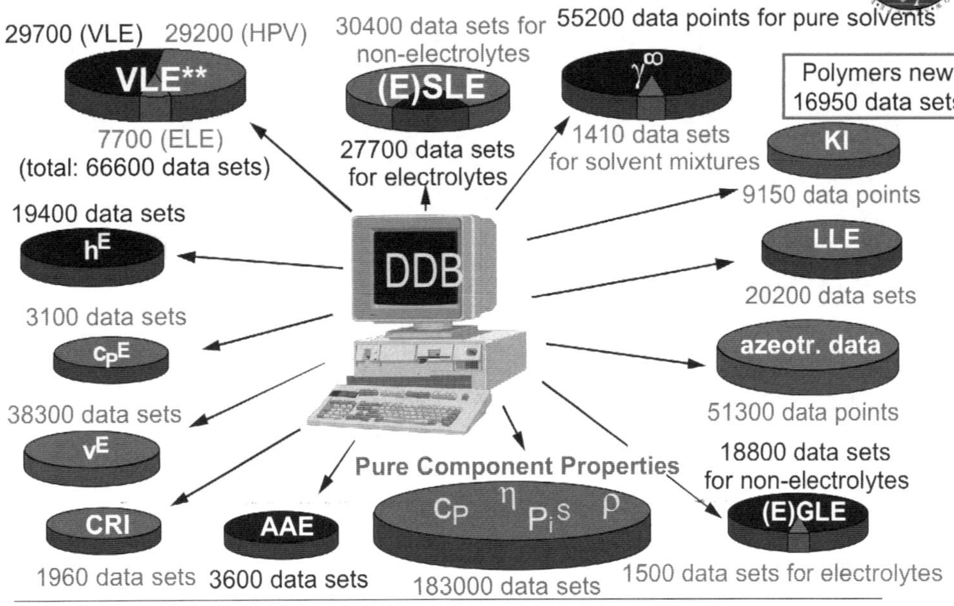

* detailed information is available via internet (www.ddbst.de)

** including unpublished VLE data from chemical companies, e.g. from the former German Democratic Republic

Different Ways to Keep the DDB Growing

- Constant further back-evaluation of older sources
- Integration of unpublished data
  - **from chemical industry**
  - **MSc. and PhD. thesis from well known research groups**
  - **These data will not be provided by online-services**
- Evaluation of new non-english scientific publications

only 10 – 15 % of the data stored in the DDB are published in book form

Digging for Data

11.02.03

Experimental and Predicted Azeotropic Data for the Quaternary System at P = 101.325 kPa Benzene (1) - Cyclohexane (2) - Acetone (3) - Ethanol (4)

	predicted (mod. UNIFAC (Do))				experimental*			
system	type of azeotrope	ϑ / °C	$y_{1,az}$	$y_{2,az}$	type of azeotrope	ϑ / °C	$y_{1,az}$	$y_{2,az}$
1-2	homPmax	77.5	0.543		homPmax	77.6	0.543	
1-3	none				none			
1-4	homPmax	68.0	0.537		homPmax	67.9	0.552	
2-3	homPmax	54.3		0.221	homPmax	53.2		0.248
2-4	homPmax	65.3		0.545	homPmax	64.8		0.553
3-4	none				none			
1-2-3	none				none			
1-2-4	homPmax	65.1	0.126	0.441	homPmax	64.9	0.113	0.462
1-3-4	none				none			
2-3-4	none				none			
1-2-3-4	none				n.a.			

* mean values of the experimental data stored in the Dortmund Data Bank

n.a.: not available

11.02.03

Selection of Selective Solvents for Extractive Distillation

Components to be separated

(1) Cyclohexane T_b = 353.86 K

(2) Benzene T_b = 353.25 K

→ **homogeneous pressure maximum azeotrope**

T_b = 351.47 K at P = 101.32 kPa

$$\alpha_{ij} = \frac{K_i}{K_j} = \frac{\gamma_i P_i^S}{\gamma_j P_j^S} = 1(or \approx 1)$$

selection criterion:

$$S_{ij,entrainer}^{\infty} = \frac{\gamma_i^{\infty}}{\gamma_j^{\infty}} >> (<<) 1$$

modified UNIFAC (Dortmund)	
selective solvent (3)	α(1,2), inf. (T [K])
Adipodinitrile	8.70 (353.56K)
2,5-Hexanedione	4.95 (353.56K)
N-Methyl-2-pyrrolidone	4.93 (353.56K)
Furfural	4.11 (353.56K)
Aniline	4.02 (353.56K)
Acetophenone	3.83 (353.56K)
Triethylene glycol	3.03 (353.56K)
Nitrobenzene	2.88 (353.56K)
Cyclohexylamine	2.80 (353.56K)
3-Methylphenol	2.10 (353.56K)
Tetrahydrofurfuryl alcohol	2.07 (353.56K)
Cyclohexanone	2.06 (353.56K)
Anisole	1.72 (353.56K)

Modified UNIFAC (Do) parameter matrix for ionic liquids

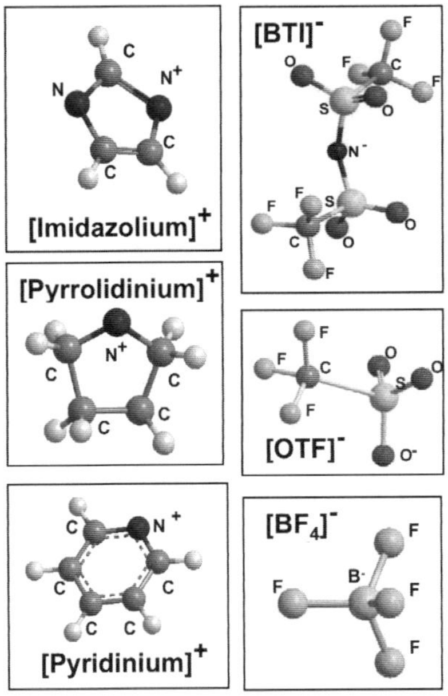

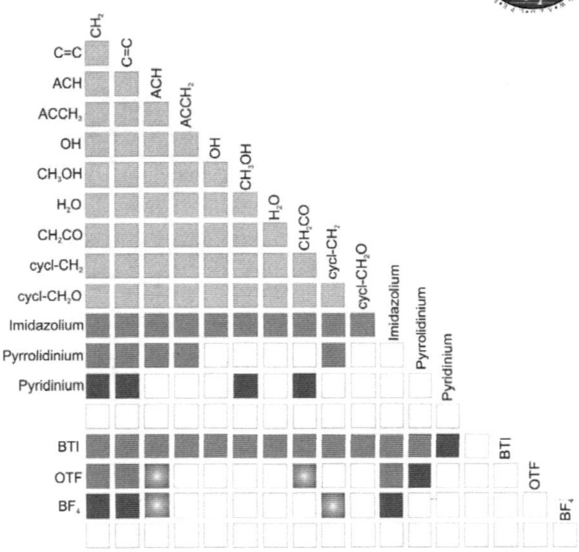

$$F = \Sigma\left(\Delta P\right)^2 + \Sigma\left(\Delta\gamma^\infty\right)^2 + \Sigma\left(\Delta h^E\right)^2$$

Experimental and predicted activity coefficients at infinite dilution for different alkanes and alkenes in 1-alkyl-3-methylimidazolium trifluoromethanesulfonate [RMIM]+[OTF]-

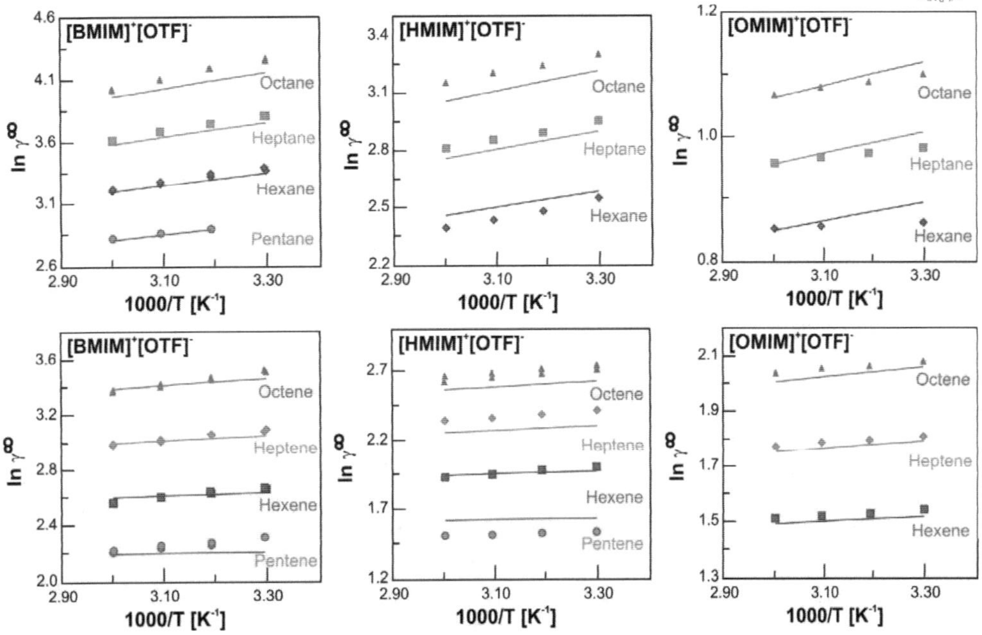

Experimental and predicted VLE data for cyclohexane – benzene with and without ionic liquid

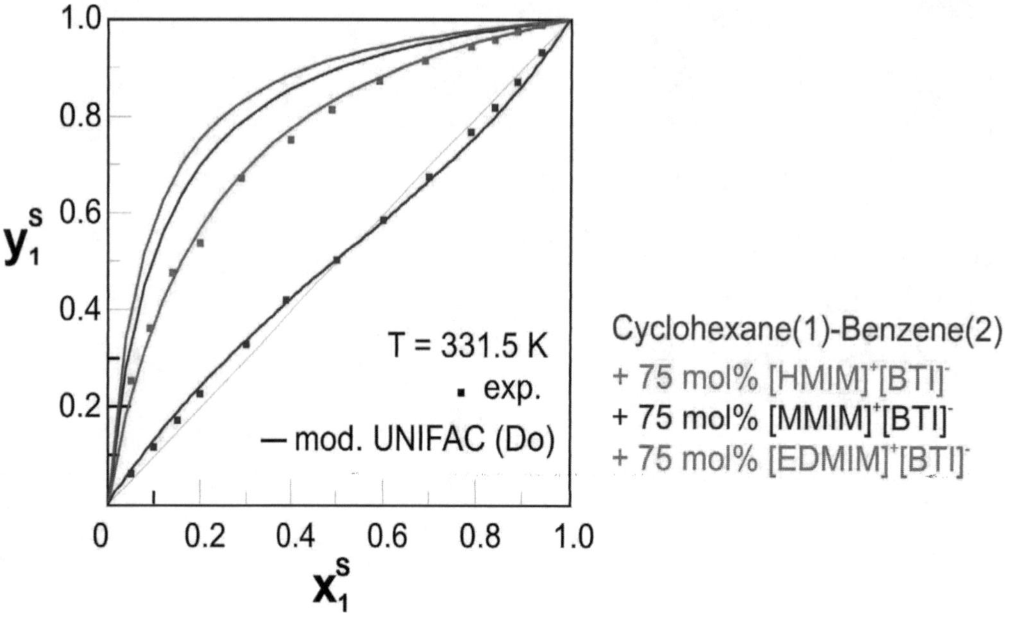

Limit of the Methods UNIFAC and Modified UNIFAC

propylene + CO + H_2 $\xrightarrow{HRh(CO)[P(C_6H_4S_3O_3Na)]_3}$ (n / i)-butyraldehyde

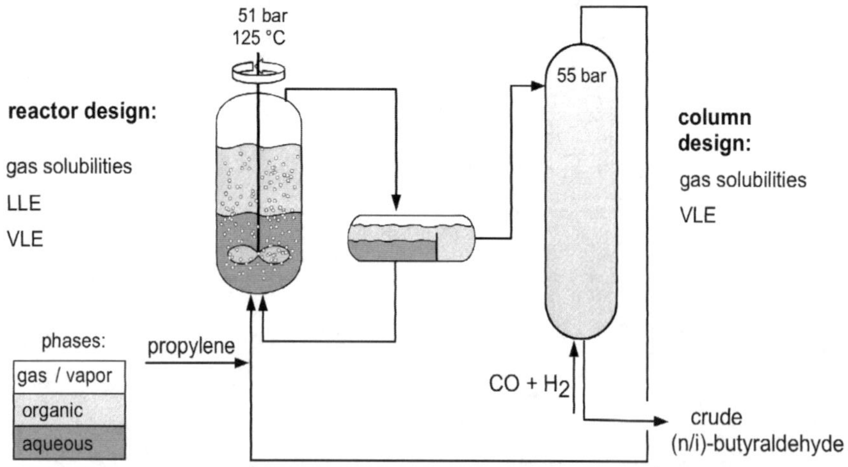

The components propylene, CO and H_2 are in the supercritical state.
=> UNIFAC and Modified UNIFAC (Dortmund) cannot be used !

05.02.0

Equations of State: van der Waals (1873)

$$P = P^{rep} + P^{att}$$

$$P^{rep} = \frac{RT}{v-b} \qquad P^{att} = -\frac{a}{v^2}$$

- a - intermolecular attraction
- b - closest packing volume

$$\left(\frac{\partial P}{\partial v}\right)_{T_{kr}} = \left(\frac{\partial^2 P}{\partial v^2}\right)_{T_{kr}} = 0$$

nobel price winner for physics 1910

$$v^3 - \left(b + \frac{RT}{P}\right)v^2 + \frac{a}{P}v - \frac{ab}{P} = 0$$

09.01.0

Otto Redlich, John M. Prausnitz, Jürgen Gmehling

$$P = \frac{RT}{v-b} - \frac{a}{T^{1/2}v(v+b)}$$

$$P = \frac{RT}{v-b} - \frac{a(T)}{v(v+b)}$$

$a(T) = a(T_c)\,\alpha(T)$

Otto Redlich, J. M. Prausnitz, ...

Giorgio Soave

05.02.03

Introduction of Improved Mixing Rules (G^E-Mixing Rules)

Acetone (1)
- Water (2)
at 60°C

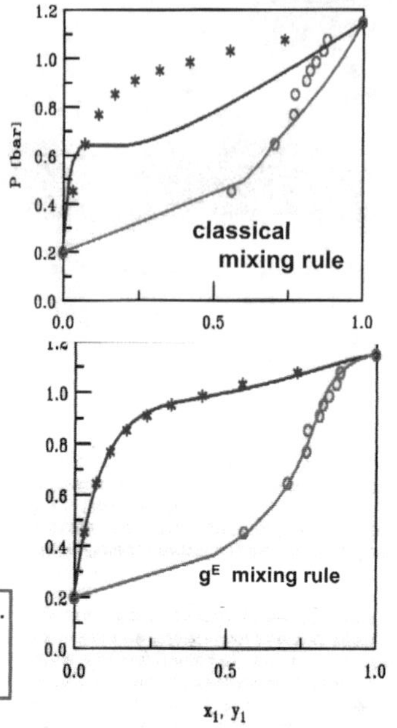

$$\frac{g^E}{RT} = \Sigma x_i \ln\gamma_i = \ln\varphi - \Sigma x_i \ln\varphi_i^*$$

$P = \infty$, g^E – model: Wilson, NRTL, ..

$$\frac{a(T)}{b} = \sum \frac{x_i a_{ii}(T)}{b_i} + \frac{g^E}{-0.6931}$$

05.02.03

PSRK (Predictive SRK Equation of State)

Huron, Vidal (1979)

P = ∞

$$\frac{a(T)}{b} = \sum \frac{x_i a_{ii}(T)}{b_i} + \frac{g^E}{-0.6931}$$

g^E-model: Wilson, NRTL, ...

PSRK (Predictive SRK)*

P_{ref} = 1 atm

$$\frac{a(T)}{b} = \sum_i \frac{x_i a_{ii}(T)}{b_i} + \frac{g^E + RT\sum_i x_i \ln\left(b/b_i\right)}{-0.64663}$$

g^E-model: original UNIFAC
Group interaction parameters:
UNIFAC matrix + parameters for 30 gases, such as CO_2, CH_4, H_2, H_2S, ..
fitted to VLE of normal and low boiling substances, gas solubilities
***Holderbaum, Gmehling (1991)**

10.02.03

Experimental and Predicted Results Using PSRK with already Available UNIFAC Parameters (VLE Data)

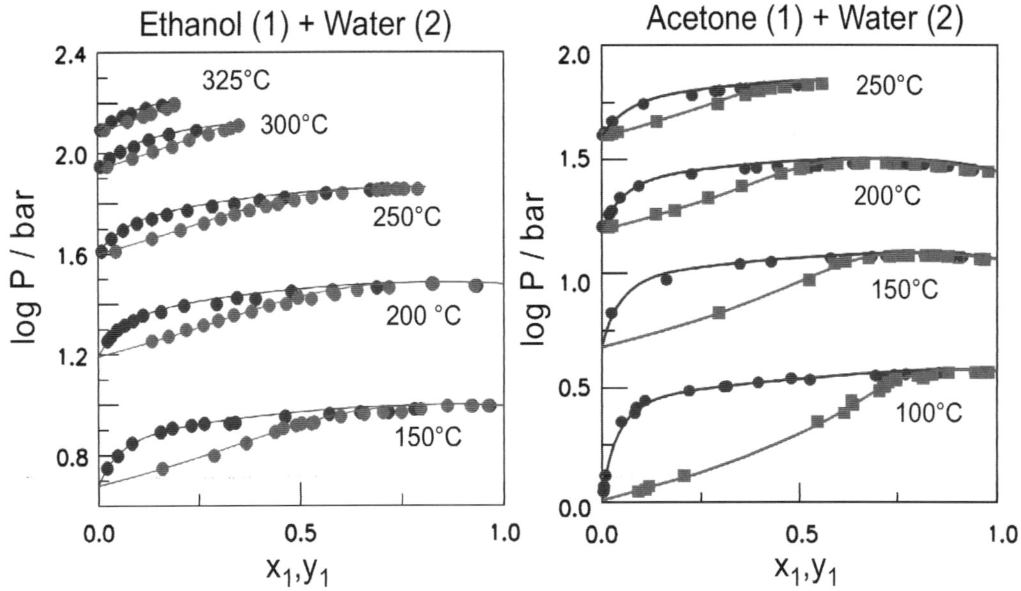

09 00 006 GC-EOS

10.02.03

Experimental and Predicted VLE Data for Different CO_2 + n-Alkane Systems Using PSRK

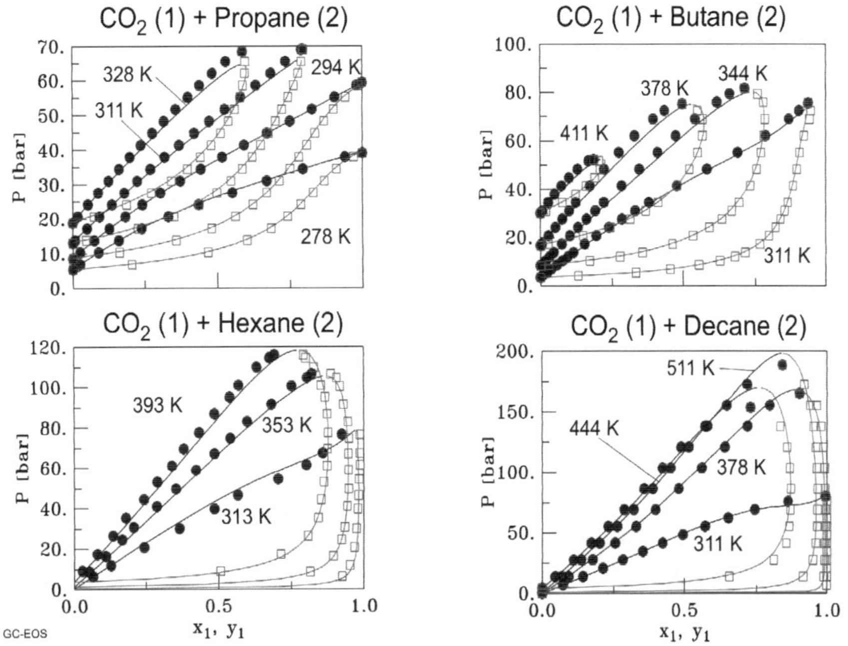

09 00 008 GC-EOS

Experimental and Predicted (PSRK) Henry Coefficients of CH_4, CO_2, COS and H_2S in Methanol

Methanreinigung.cdr

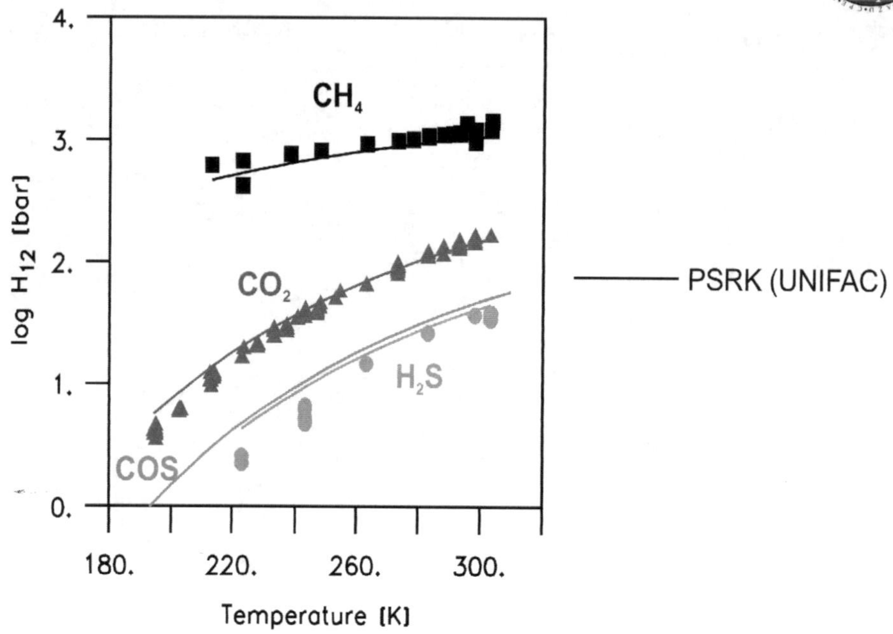

Experimental and Calculated Gas Solubilities (LIFAC – PSRK)

$$\ln \gamma_i = \ln \gamma_i^{LR} + \ln \gamma_i^{MR} + \ln \gamma_i^{SR}$$

long-range (Debye-Hückel) — middle-range (Pitzer) — short-range (UNIFAC)

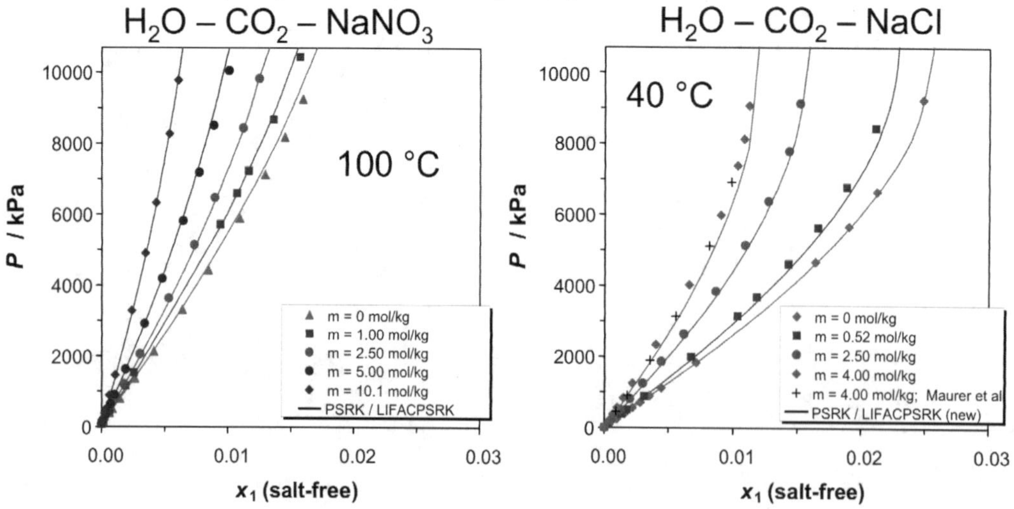

Conclusion

- great progress has been achieved for predictive thermodynamic models
- a comprehensive data bank was a prerequisite for this development

Predictive model	Data used for fitting the parameters
UNIFAC	VLE (LLE)
Modified UNIFAC	VLE (LLE), azeotropic data, h^E, γ^∞, SLE of eutectic systems
LIQUAC (LIFAC)	VLE of electrolyte systems, osmotic coefficients, mean activity coefficients, salt solubilities
Group contribution EOS PSRK	VLE of normal boiling and low boiling substances, gas solubilities

Recent development: improved group contribution equation of state
VTPR (volume-translated Peng-Robinson)

CONTEXT OF THE CONTRIBUTOR

Research

Industrial Chemistry – University of Oldenburg
Prof. Dr. J. Gmehling - Research Projects, Model Development

UNIFAC Consortium
Mixture Property Estimation

TRU – Durban, SA
Pure Component Property Estimation

CAT - Center for Applied Thermodynamics

Products

DDBST
Data Banks, Software, Consulting

LTP (Associated Institute)
Experimental Services, Consulting

Education

Graduate Courses Thermodynamics, Training Courses DDB/DDBSP
Text books: Industrial Chemistry, Unit Operations, Thermodynamics,

Sales, Marketing, Custom Tailored Products

DECHEMA
Frankfurt, Germany

FIZ CHEMIE
Berlin, Germany

EPCON Int.
Houston, Texas

Mitsubishi Chemical
Kurashiki, Japan

Companies founded

LTP GmbH

www.ltp-oldenburg.de

DDBST GmbH

www.ddbst.de

Members of the UNIFAC Consortium (2009)

The UNIFAC Consortium

Uhde ThyssenKrupp
سابك sabic
gsk GlaxoSmithKline
processium
ARKEMA
sasol reaching new frontiers
OXEA
Linde
LYONDELL
Givaudan
REPSOL YPF
DDBST DORTMUND DATA BANK SOFTWARE & SEPARATION TECHNOLOGY
CHEMSTATIONS
ULB UNIVERSITÉ LIBRE DE BRUXELLES, UNIVERSITÉ D'EUROPE
mg engineering Lurgi
AkzoNobel Tomorrow's Answers Today
EVONIK INDUSTRIES
Rhodia
SIEMENS
EPCON International
DOW
REFINE
Roche
UNIVERSITY OF KWAZULU-NATAL
ProSim
QVF
DUPONT
Mitsui Chemicals 新しい化学の創造
DSM Chemicals & Materials
ifp
Laboratory for Thermophysical Properties GmbH
bp
NESTE OIL
ExxonMobil Research and Engineering
cognis
Lonza group
Ciba
PURAC
WACKER
SOLVAY
EASTMAN Eastman Chemical Company
BASF The Chemical Company
MITSUBISHI CHEMICAL
Bayer Technology Services
SUMITOMO CHEMICAL
TU Graz

UNIFAC Sponsor-Meeting 2008

Parameter Matrix for Modified UNIFAC (Dortmund)

New Structural Groups:

46 cy-CONC
47 CONR
48 $CONR_2$
49 HCONR
50 ACCN
51 NCO
52 ACS
53 Epoxy
54 Anhydrides
55 Carbonates
56 Sulfones
57 ACCHO
58 ACCOOH
59 ACCOO
60 OCCOH
61 CH_2S
62 Lactames
63 Lactones
64 Peroxides
74 Acetals
75 ACNR2
76 ACNHR
77 Furan
78 cy-CNH
79 cy-CNR
80 SiH
81 SiO
82 Oxime
83 ACCO

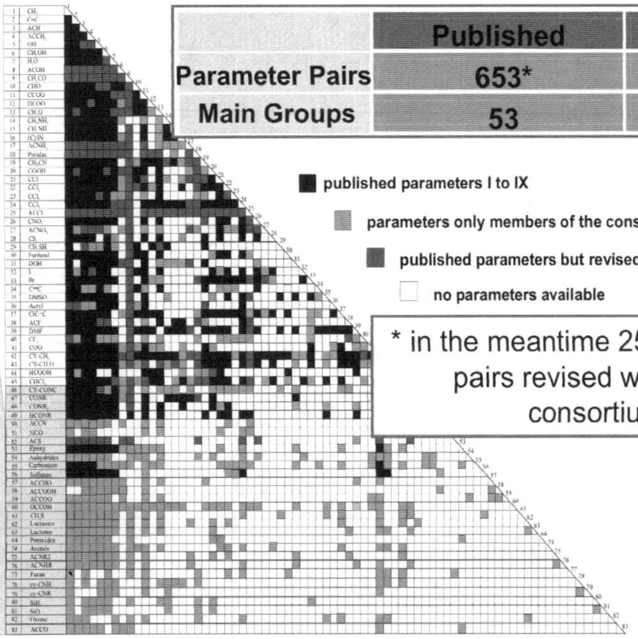

	Published	Consortium
Parameter Pairs	653*	1332
Main Groups	53	87

- published parameters I to IX
- parameters only members of the consortium
- published parameters but revised within the consortium
- no parameters available

* in the meantime 255 parameter pairs revised within the consortium

Deviations Between Experimental and Calculated VLE (Data Base: 1358 Consistent Data Sets)

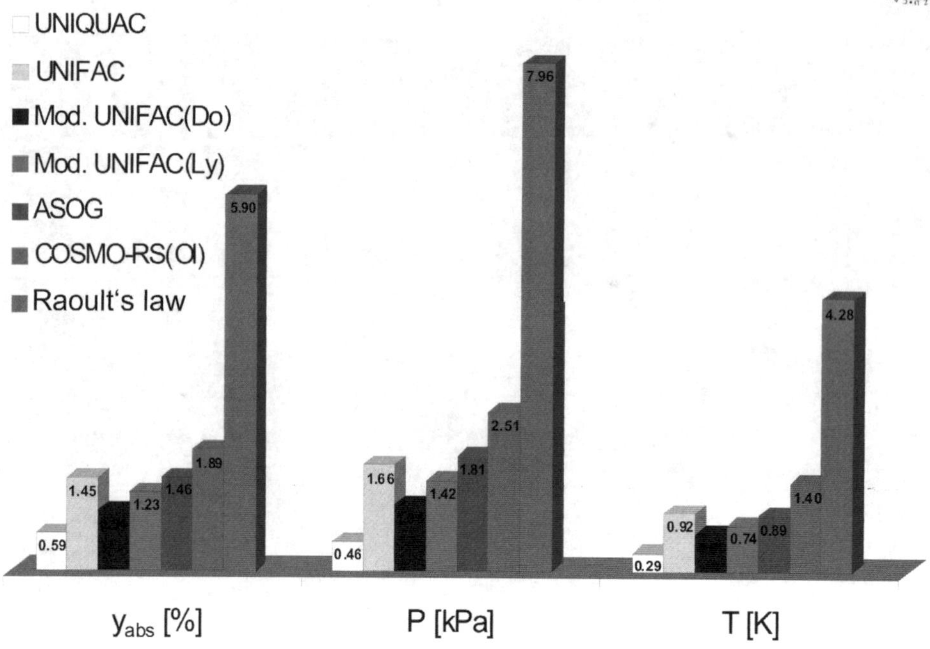

UNIFAC Parameter Matrix

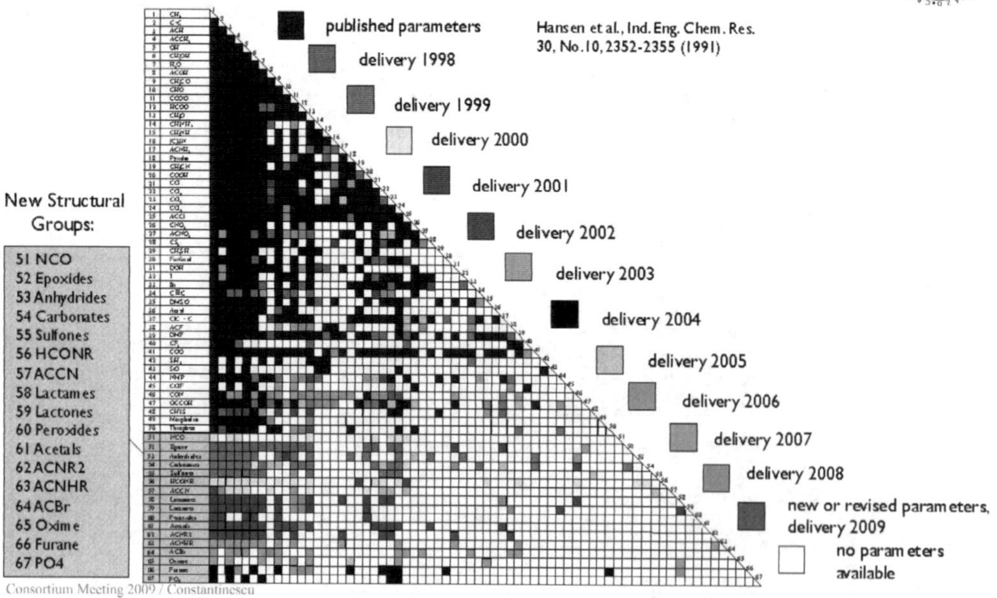

Consortium Meeting 2009 / Constantinescu

PSRK Matrix

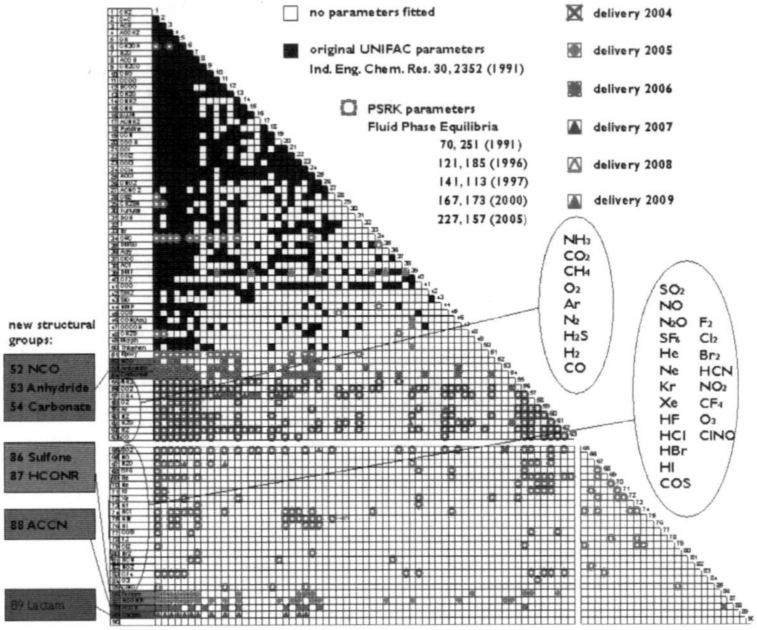

Consortium Meeting 2009 / Constantinescu

Summary

Group interaction parameters:

method	parameter pairs			
	New in 2009	revised	published	consortium
UNIFAC	6	--	635 50 main groups	985 67 main groups
Modified UNIFAC	39	255	653 53 main groups	1332 87 main groups
Modified UNIFAC (COSMO-RS(Ol))	25	--	8	106
PSRK	6	1	955 82 main groups	1086 89 main groups

Mod. UNIFAC Matrix - Gaps filled with COSMO-RS (Ol)

Measurement difficult (highly reactive, explosive or toxic components): Prediction with the quantum-mechanical method COSMO

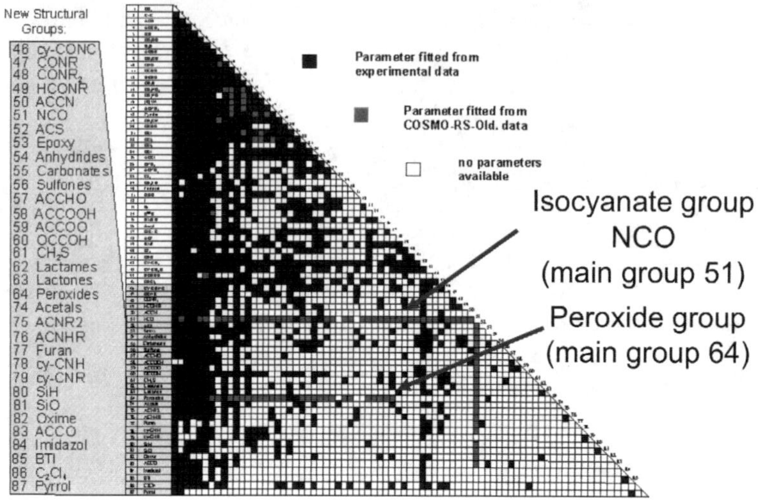

Overview Delivery 2009

Experimental Delivery 2009

	2009	Total
Vapor-Liquid Equilibrium Data	20	231
Heats of Mixing Data	23	257
Solid-Liquid Equilibrium Data		13
Azeotropic Data		3
Gas-Liquid Data	12	27
Activity Coefficients at infinite dilution	24	28

Experimental Facilities

vapor-liquid equilibria (VLE)

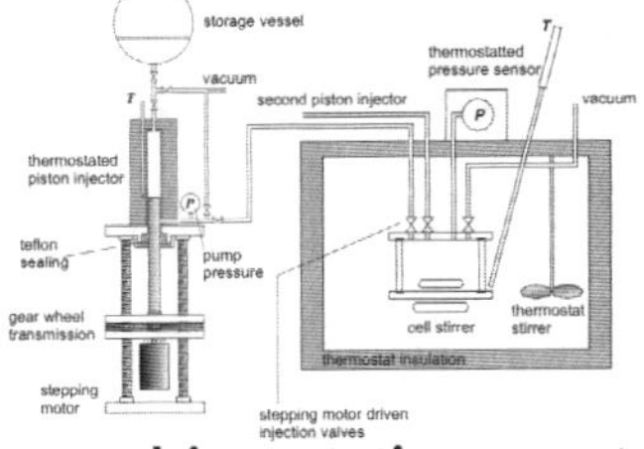

computer-driven static apparatus

excess enthalpies (h^E)

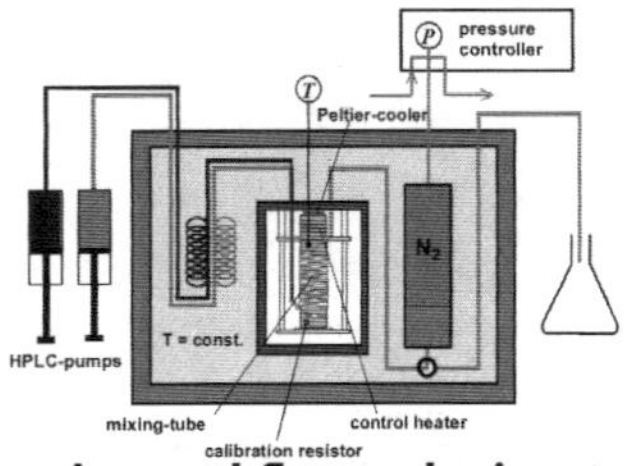

isothermal flow calorimetry

azeotropic data (AZD)

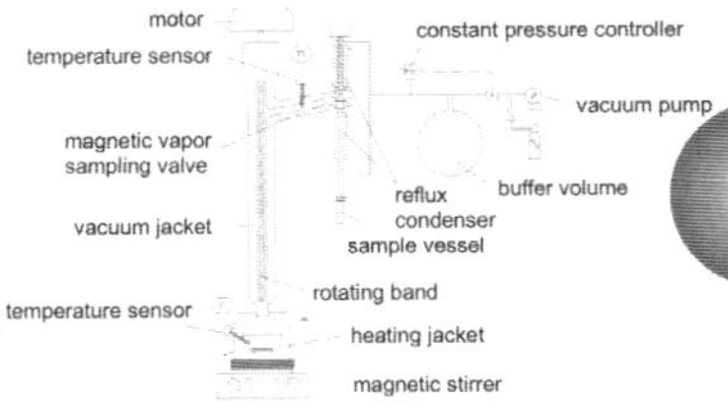

wire band column

activity coefficients at infinite dilution (ACT)

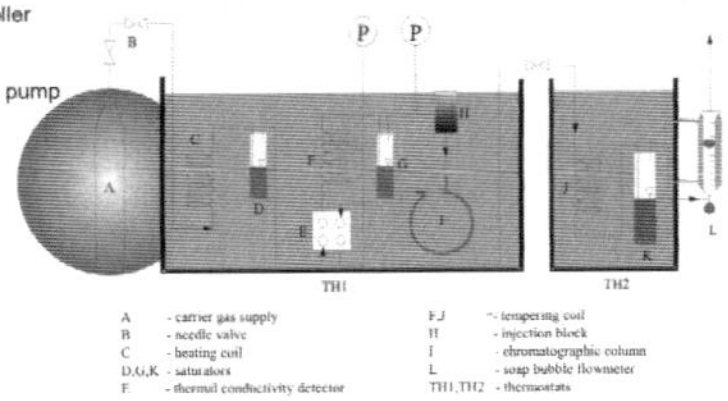

GLC

solid-liquid equilibria (SLE)

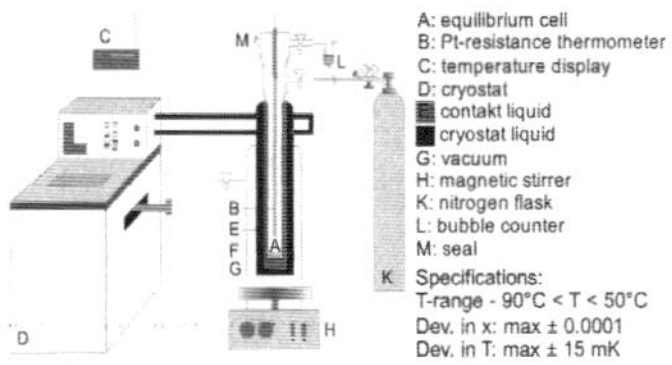

visual method

Consortium Meeting 2009 / Constantinescu

Overview Delivery 2009

- systematic measurements for main groups:
  11 „CCOO“ (acetate group)
  12 „HCOO“ (formide group)
  78 „cy-C-NH“ (cyclic amine group)

- two new main groups in Modified UNIFAC (Dortmund):
  86 „C_2Cl_4“ (Tetrachloroethylene)
  87 „$C_4H_9N^+$ (Pyrrol)

(+ 5 new subgroups)

3 new subgroups in UNIFAC

Consortium Meeting 2009 / Constantinescu

Objectives 2009/2010

Revision and Extension of current Parameter Matrices for UNIFAC, Mod. UNIFAC

- Filling Gaps
- New Main Group for acroleine
- Subgroup upgrading

Definition of new Main Groups for PSRK

- adapt PSRK matrix to UNIFAC

Filling Gaps in the Modified UNIFAC Matrix with COSMO-RS (OI)

- systems where no experimental data are available

Experimental work:

- Determination of Phase Equilibrium Data for the Revision of Interaction Parameters
- Systematic Measurements (VLE, AZD, hE, ACT, SLE, GLE) for filling Gaps

UNIFAC Consortium Team (Sept. 2009)

Dr. Dana Constantinescu

(UNIFAC, mod. UNIFAC, management consortium, customization files, ..)

Denis Krotov

(PSRK)

Experimental work:

VLE, GLE, h^E, SLE, γ^{∞}, azeotropic data

Rainer Bölts

Helmut Stojek

Kathrin Schreiber

most of them working part time on the project

Acknowledgement

Prof. Gmehling would like to thank

Prof. Dr. Ulfert Onken, Dr. Hermann Stage, Prof. Dr. Aage Fredenslund, Prof. Dr. Jiding Li, Prof. Dr. Jian Chen, Prof. Dr. Weidong Yan, Prof. Dr. John M. Prausnitz, Dr. Antje Jakob

all PhD students, in particular:

Jürgen Rarey, Jochen Menke, Wolfgang Arlt, Ulrich Weidlich, Bärbel Kolbe, Hans-Martin Polka, Thomas Holderbaum, Martin Schiller, Kai Fischer, Jens Ahlers

the long standing technical co-workers Rainer Bölts, Bernd Werner

a large number of foreign guests

and the different institutions:

Deutsche Forschungsgemeinschaft (DFG), the Ministry of Research (BMFT), AiF, DDBST team, members of the UNIFAC consortium

I thank Todd Willman for the inventation to this conference and you for your attention and patience !

Data published by E. Sapei, P. Uusi-Kyyny. K. I. Keskinen, J. Aittamaa, Fluid Phase Equilib. 279, 81-86 (2009)

—— modified UNIFAC

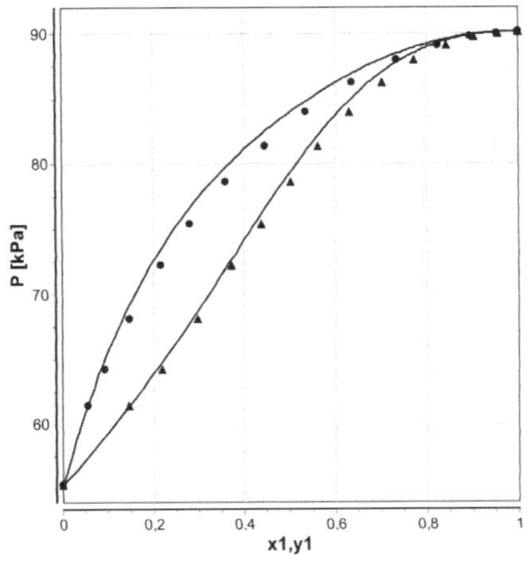

2,2,4-trimethylpentane (1) – 3-methylthiophene (2) at 368.15 K

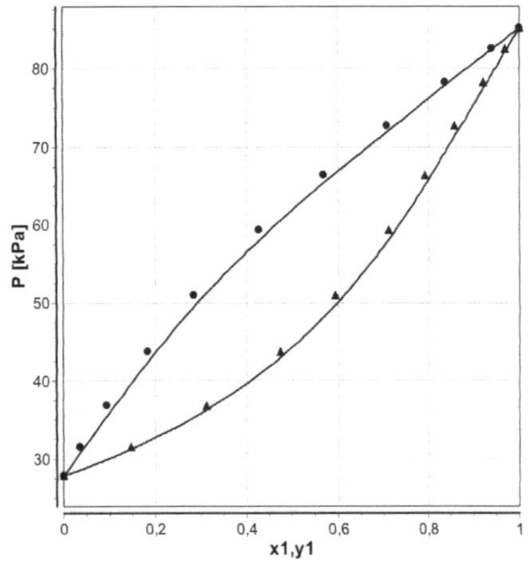

cyclohexane (1) - 3-methylthiophene (2) at 348.15 K

Data published by E. Sapei, P. Uusi-Kyyny. K. I. Keskinen, J. Aittamaa, Fluid Phase Equilib. 279, 81-86 (2009)

—— modified UNIFAC

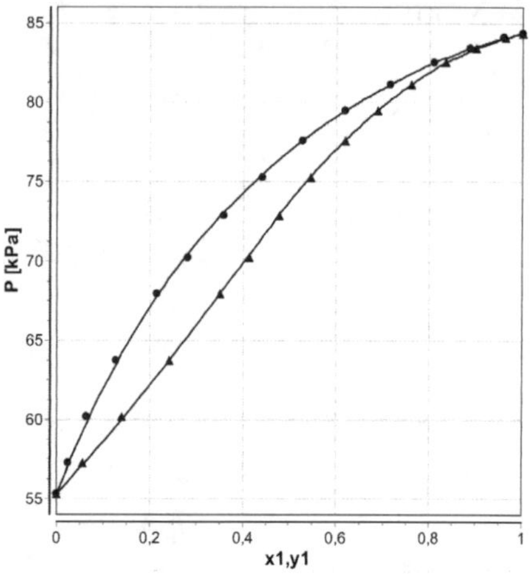

2,4,4-trimethyl-1-pentene (1) – 3-methylthiophene (2) at 368.15 K

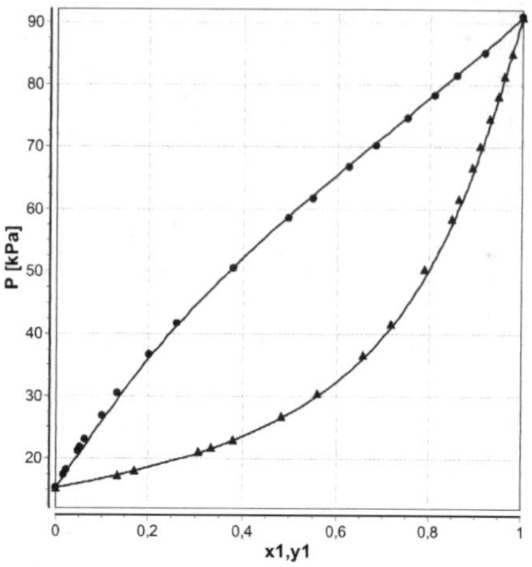

1-hexene (1) - 3-methylthiophene (2) at 333.15 K

Results predicted with mod. UNIFAC(Do) for h^E Data by using new Parameters - Measurements performed in our laboratory

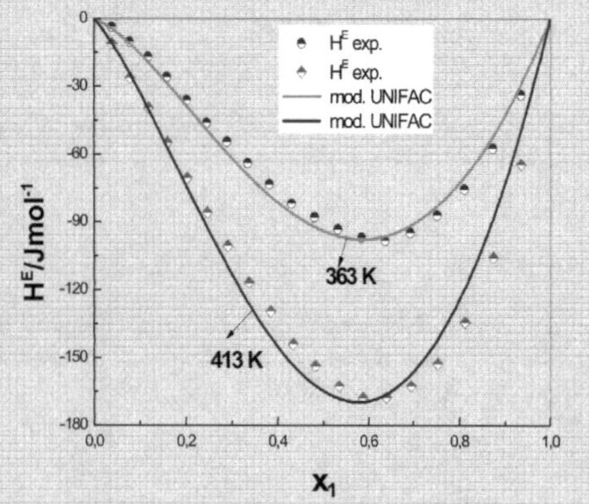

Diethyl ether(1) + Thiophene(2)
Interaction parameters between main groups 13/52 „CH_2O" / „ACS"

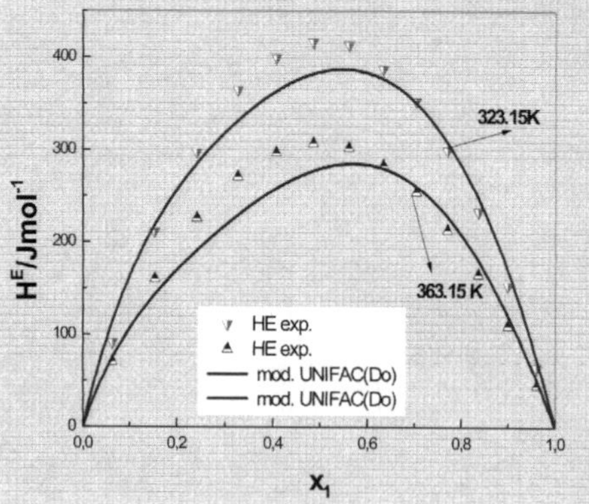

Pyrrolidine(1) + Toluene (2)
Interaction parameters between main groups 78/4 „cy-C-NH" / „$ACCH_2$"

Consortium Meeting 2009 / Constantinescu

Results predicted with the new added main group 86 (C_2Cl_4) in Mod. UNIFAC (Dortmund)

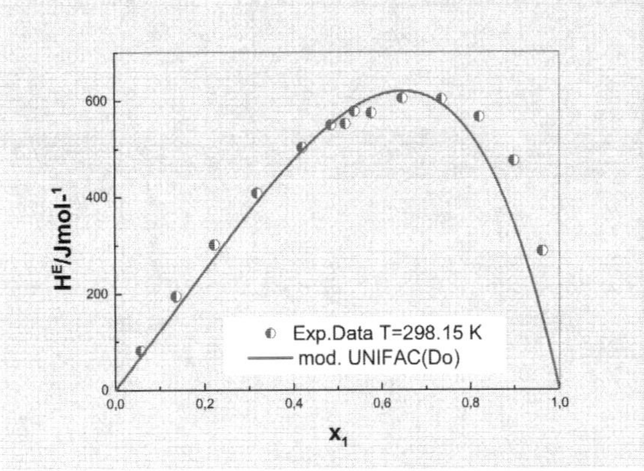

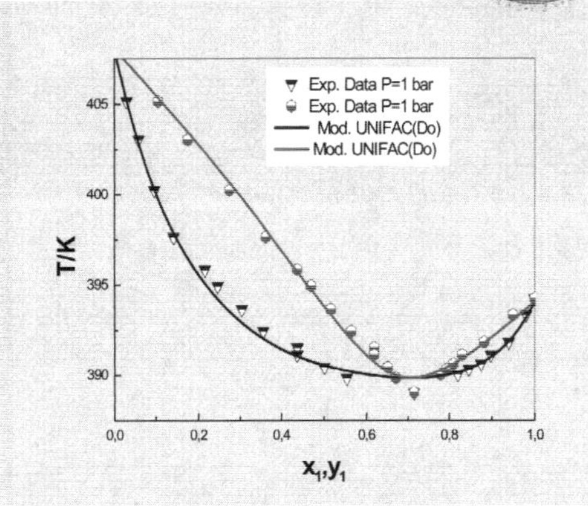

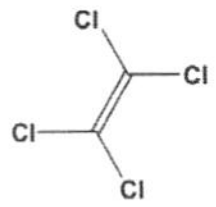

Tetrachloroethylene(1) + 2-Ethoxyethanol(2)
Interaction parameters between main groups 86/60 „C2Cl4" / „OCCOH"

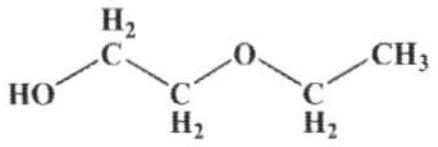

Consortium Meeting 2009 / Constantinescu

Results predicted with the new added main group 86 (C_2Cl_4) in Mod. UNIFAC (Dortmund

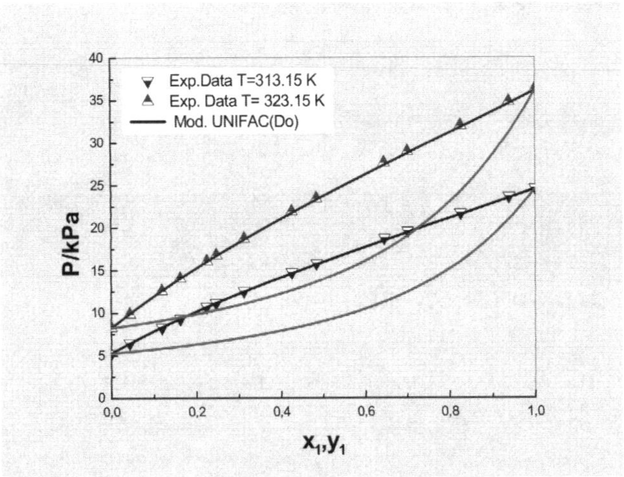

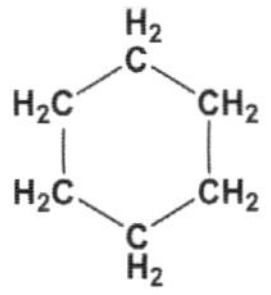

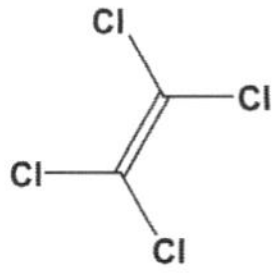

Cyclohexane (1) + Tetrachloroethylene (2)
Interaction parameters between main groups 42/86 „ cy-CH_2" / „ C_2Cl_4"

Consortium Meeting 2009 / Constantinescu

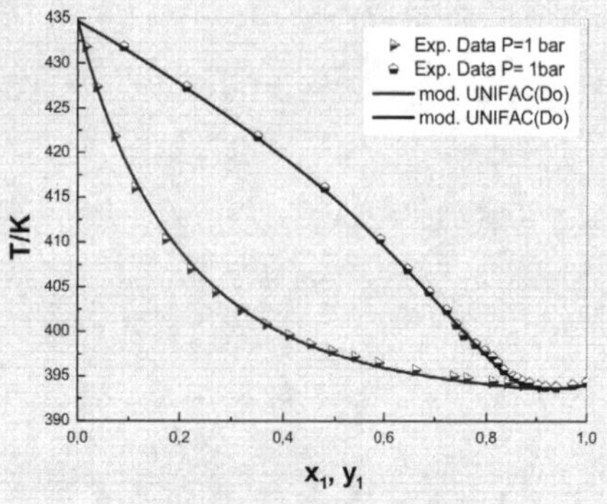

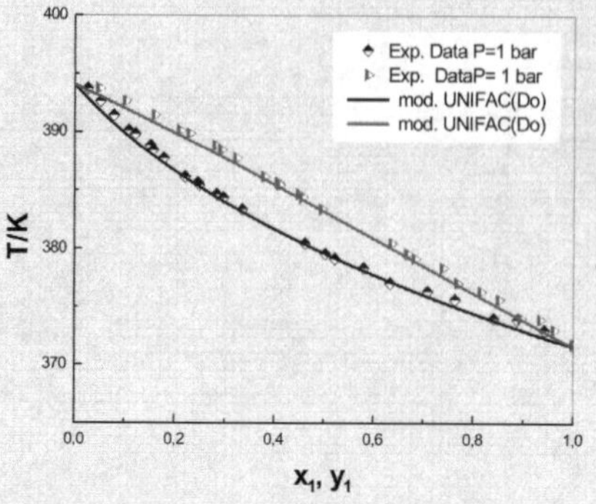

Tetrachloroethylene (1) + Furfural (2)
Interaction parameters between main groups 86/30 „ C_2Cl_4“ / „ Furfural“

Heptane (1) + Tetrachloroethylene (2)
Interaction parameters between main groups 1/86 „ CH_2“ / „ C_2Cl_4 “

Consortium Meeting 2009 / Constantinescu

Weaknesses of group contribution methods

Isomer effects can not be described correctly

(but the influence of the standard fugacity P_i^s, P_i^{subl} (Δh_m, T_m) is more important), e.g. in the case of VLE, SLE, ..
no improvement is achieved with other approaches, e.g. COSMO-RS

Problems exist, when a large number of functional groups is present (proximity effects)

Poor γ^∞-results and LLE for water with alkanes (alkenes) or naphthenes

Number of h^E-Data Sets Stored in the DDB as a Function of Temperature

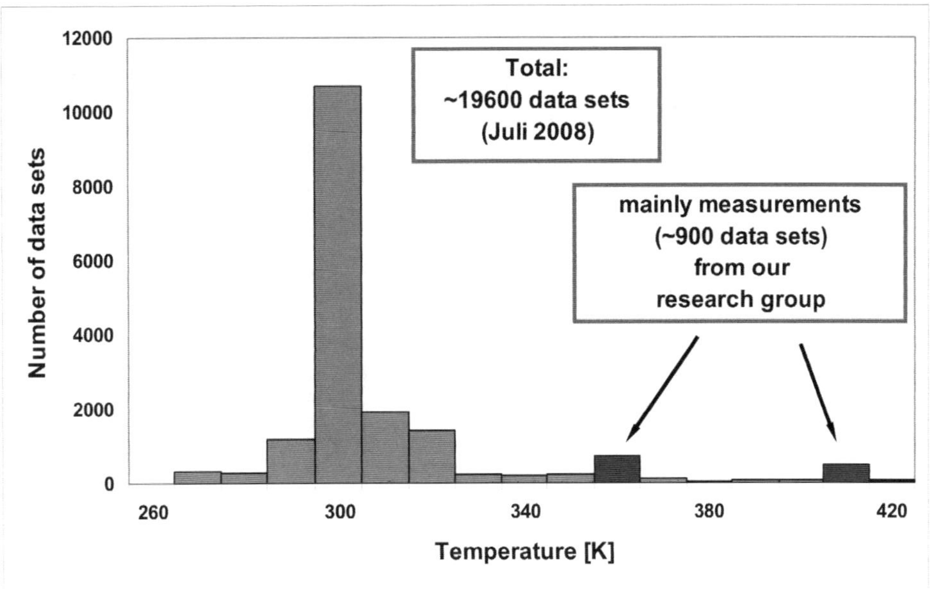

Experimental and predicted activity coefficients at infinite dilution of different alkanes in alkylmethylimidazolium-bistrifluorosulfonylimide $[RMIM]^+[BTI]^-$

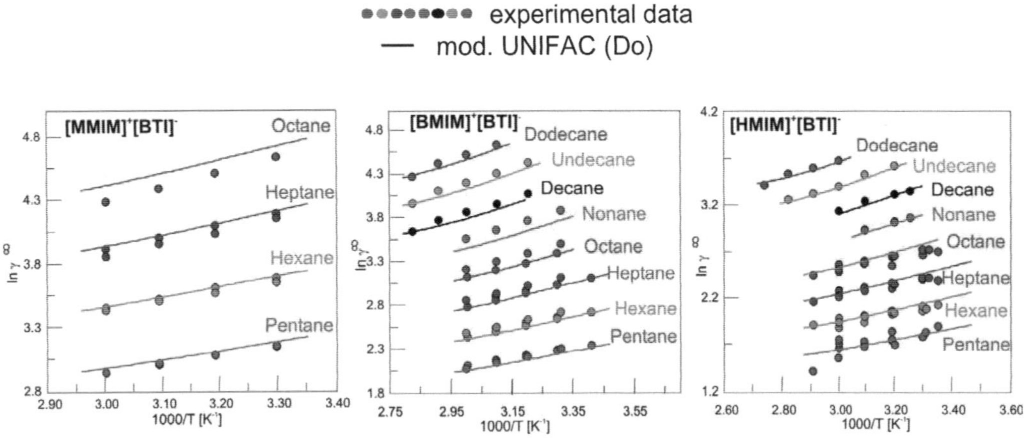

18.03.99

The LIFAC (LIQUAC)-Model

$$\ln\gamma_i = \ln\gamma_i^{LR} + \ln\gamma_i^{MR} + \ln\gamma_i^{SR}$$

Long Range	**Middle Range**	**Short Range**
Debye-Hückel equation for solvent mixtures	**empirical version of the Pitzer approach**	**UNIFAC (UNIQUAC)**
no further parameters required	**required parameters b_0, b_1**	**original UNIFAC parameters**

b_o and b_1 fitted simultaneously to VLE of electrolyte systems, mean activity coefficients $\gamma_\pm$ and osmotic coefficients Φ

LIFAC Parameter Matrix

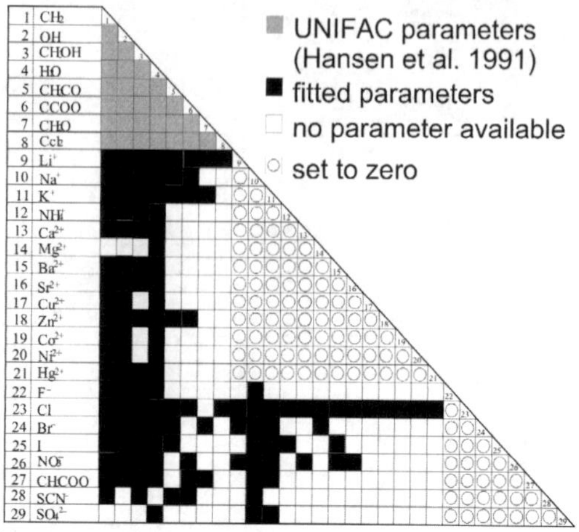

53

Solubility and Mean Activity Coefficients of Different Salts in Water

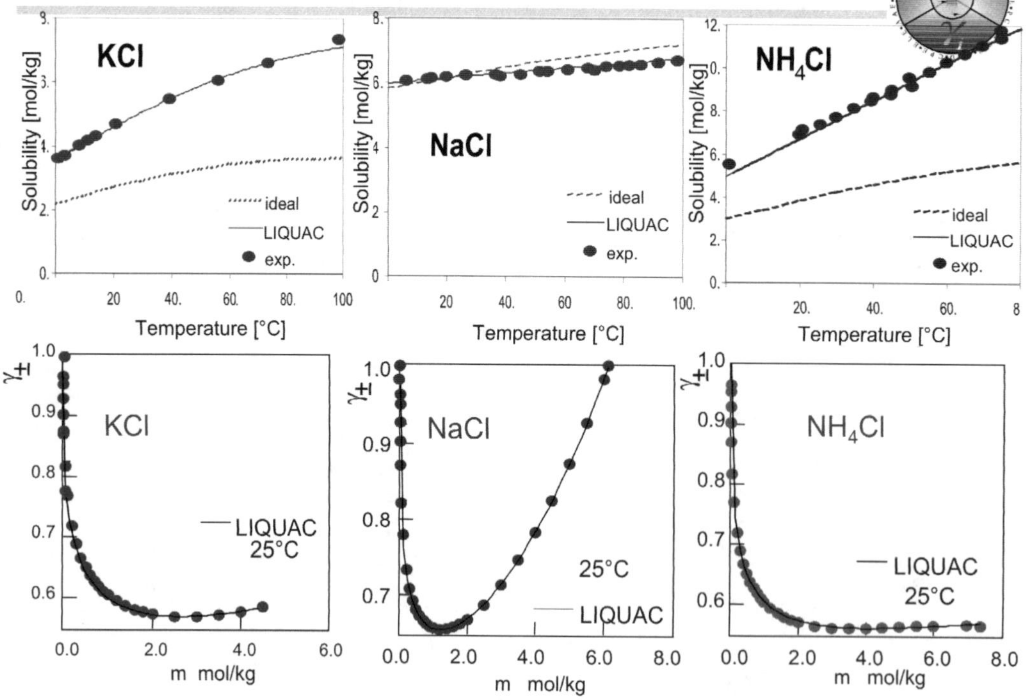

Experimental and Predicted Solubility (LIQUAC) of KCl in the System Water(1)-Methanol(2)

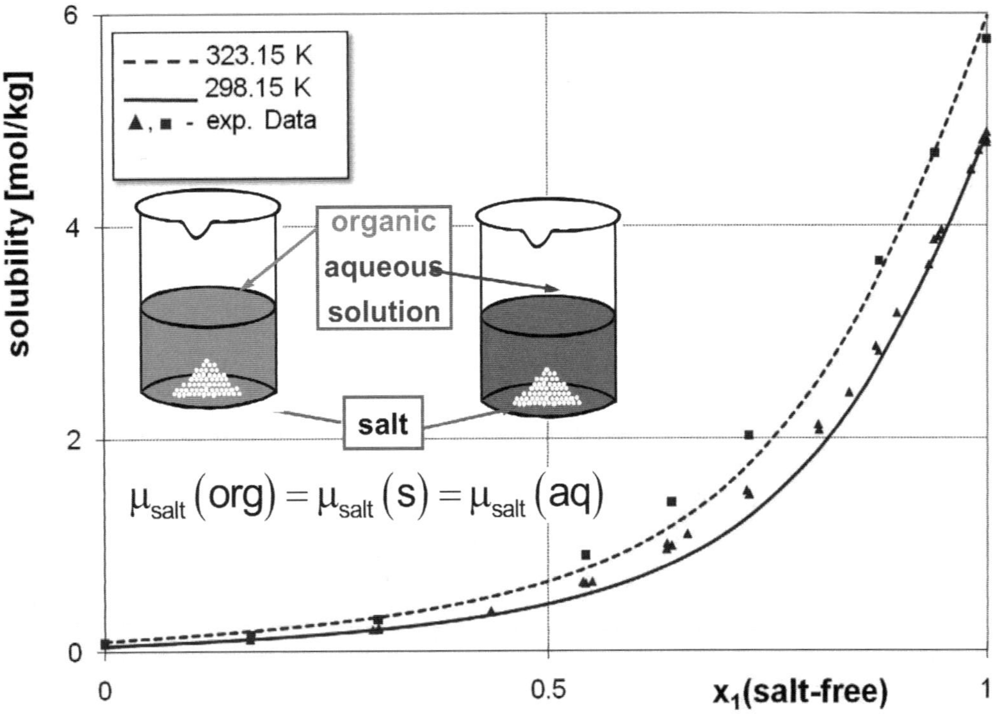

The BP Baker Report: Assessing its Impact on Process Safety Culture

Stephanie C. Payne Jennifer M. Rodriguez
Mindy E. Bergman

Texas A&M University

The Mary Kay O'Connor Process Safety Center supported this research

Stephanie C. Payne

- Associate Professor, Texas A&M University
- Faculty Fellow, Mary Kay O'Connor Process Safety Center
- Ray A. Rothrock Fellow
- Ph.D. 2000 Industrial/Organizational Psychology, George Mason University

Current Study Objectives

1. Gather information from industry insiders about the impact of the BP Baker Report on safety

2. Determine whether the BP Process Safety Culture Survey is becoming an industry benchmarking tool

BP Texas City Refinery Explosion

- March 2005 - 15 killed, 170 injured
- Chemical Safety and Hazard Investigation Board recommended independent panel assessment
  - Focus on process safety rather than occupational safety
  - Review, interviews, survey, etc.
  - BP Baker Panel Report issued Jan 2007

BP Baker Report Findings

- Personal vs. process safety
  - Distinction
  - Indicators not the same nor substitutable
- Importance of safety culture
- Importance of integrating contractors

Method

- February 2009
- Mary Kay O'Connor Process Safety Center listserv subscribers invited to participate
- 379 respondents
- 15-minute online anonymous survey
- Five $50 cash prizes awarded via lottery

Participants

- Sex: 77% male
- Age: 21-83
  M=51.18, *SD*=10.78
- Race:
  70% Caucasian
- Tenure:
  M=23.94; *SD*=11.45

Job titles

Internal

- Engineers (*n*=90)
- Safety mgrs/dirs (*n*=55)
- Safety personnel (*n*=37
- Ops mgrs (*n*=25)
- Process safety personnel/mgrs (*n*= 24)

External

- Consultants (*n*=60)
- Gov't ees (*n*=16)

Reading of BP Baker Report

- 86% read at least some of the report

Reasons for reading

(check all that apply; *n* = 327)

- 21% required to read as a part of their job
- 15% supervisor asked opinion of it
- 31% presented info from it to others
- 76% personally interested in reading it

Lessons Learned from the Report

<u>Open-ended responses</u> (*n* = 188)

- Management should give priority to safety (24%)
- Difference between personal and process safety (10%)
- Relationship between culture and safety (10%)

Importance of Report (*M* = 4.12)

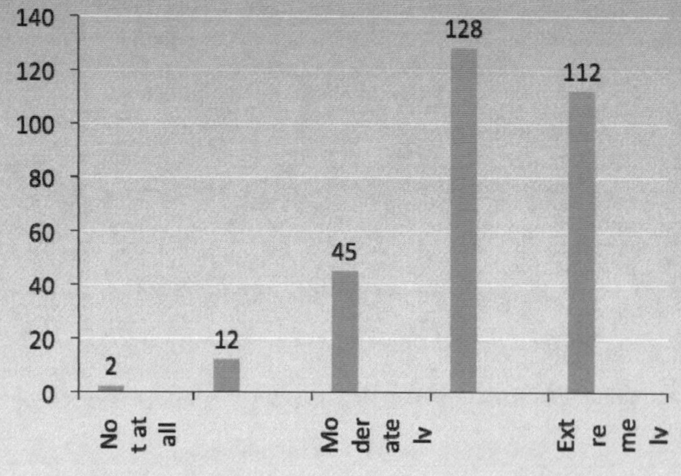

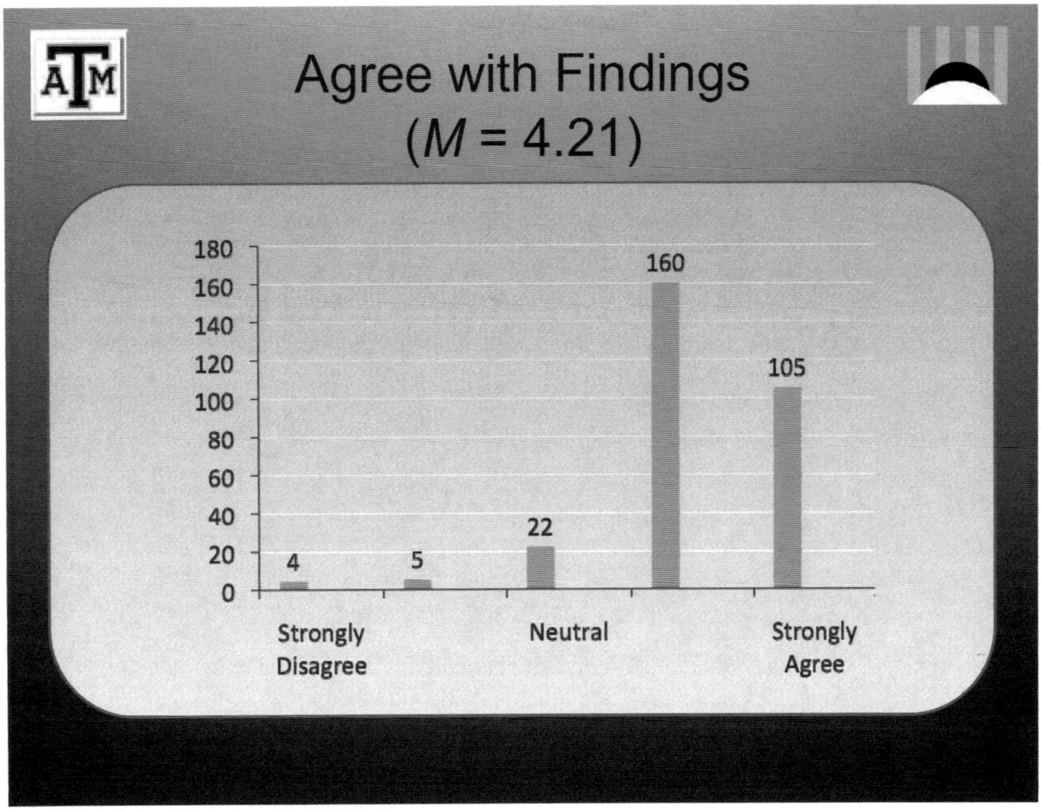
Agree with Findings
(M = 4.21)
180
160
140
120
100
80
60
40
20
0
4
5
22
160
105
Strongly Disagree
Neutral
Strongly Agree

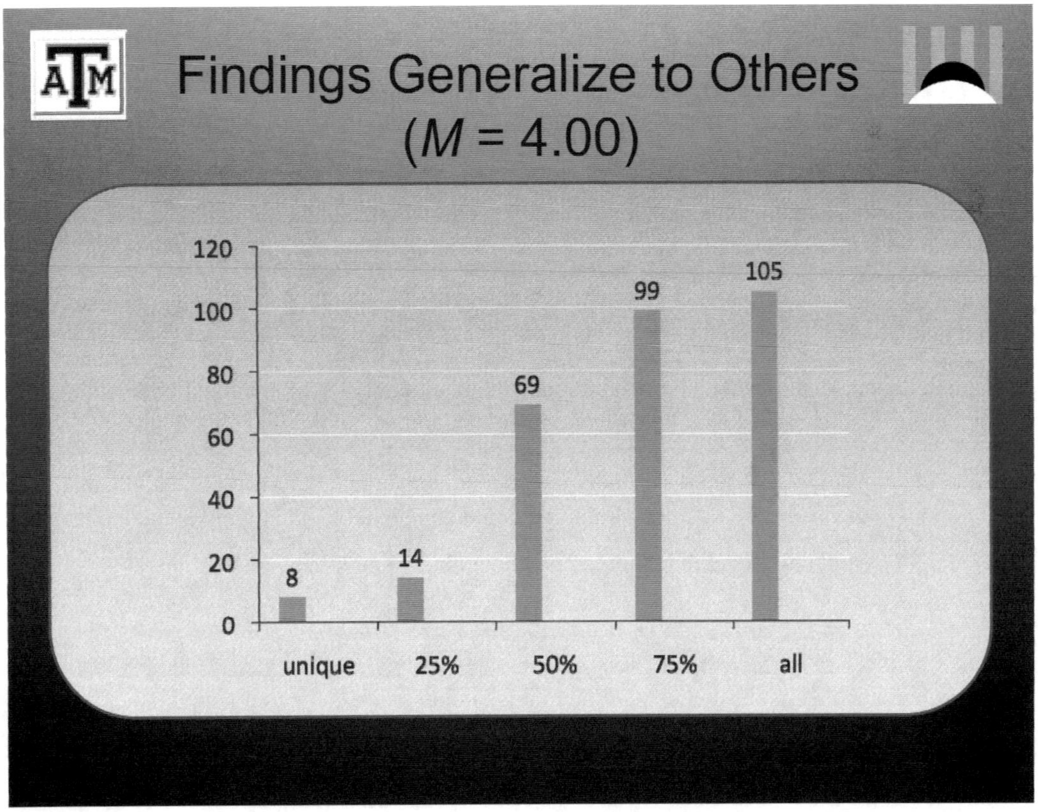
Findings Generalize to Others
(M = 4.00)
120
100
80
60
40
20
0
8
14
69
99
105
unique
25%
50%
75%
all

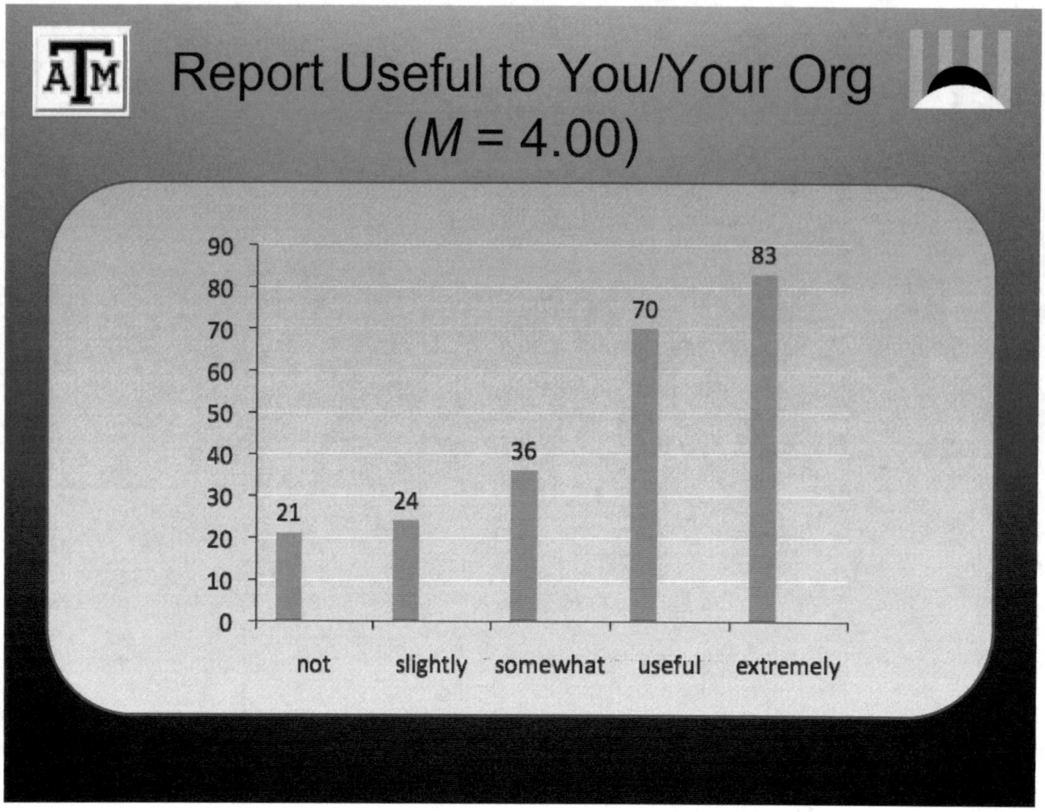
Report Useful to You/Your Org
(M = 4.00)
90
80
70
60
50
40
30
20
10
0
21
24
36
70
83
not
slightly
somewhat
useful
extremely

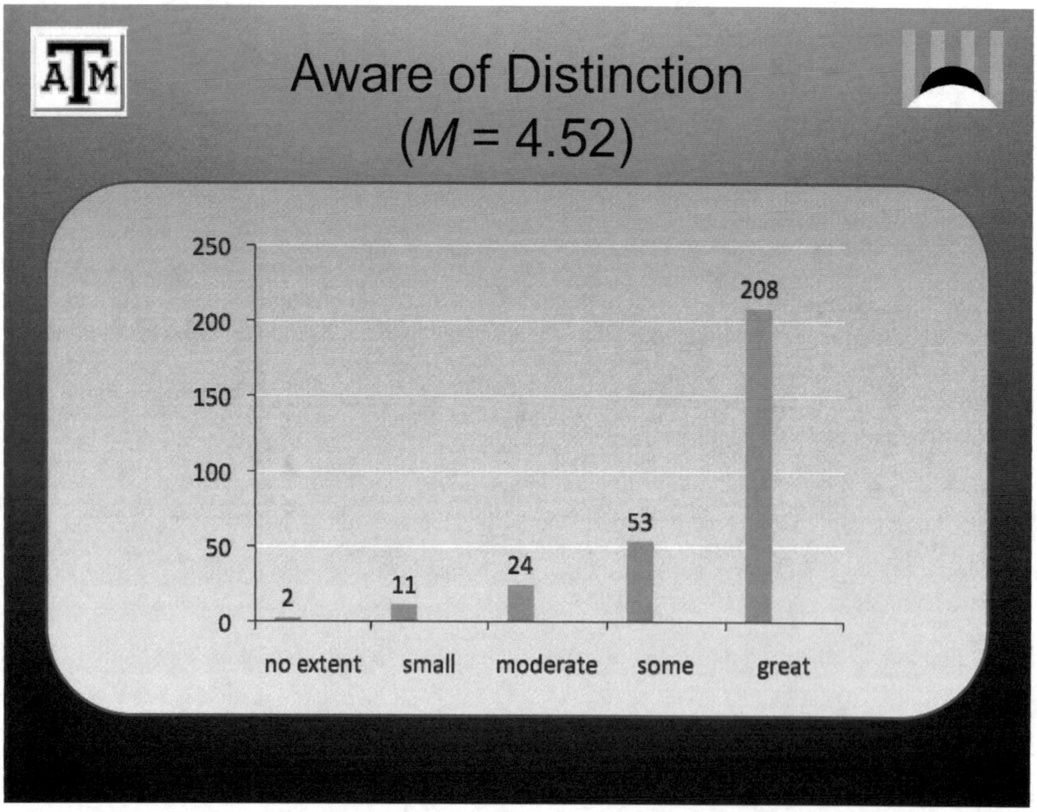
Aware of Distinction
(M = 4.52)
250
200
150
100
50
0
2
11
24
53
208
no extent
small
moderate
some
great

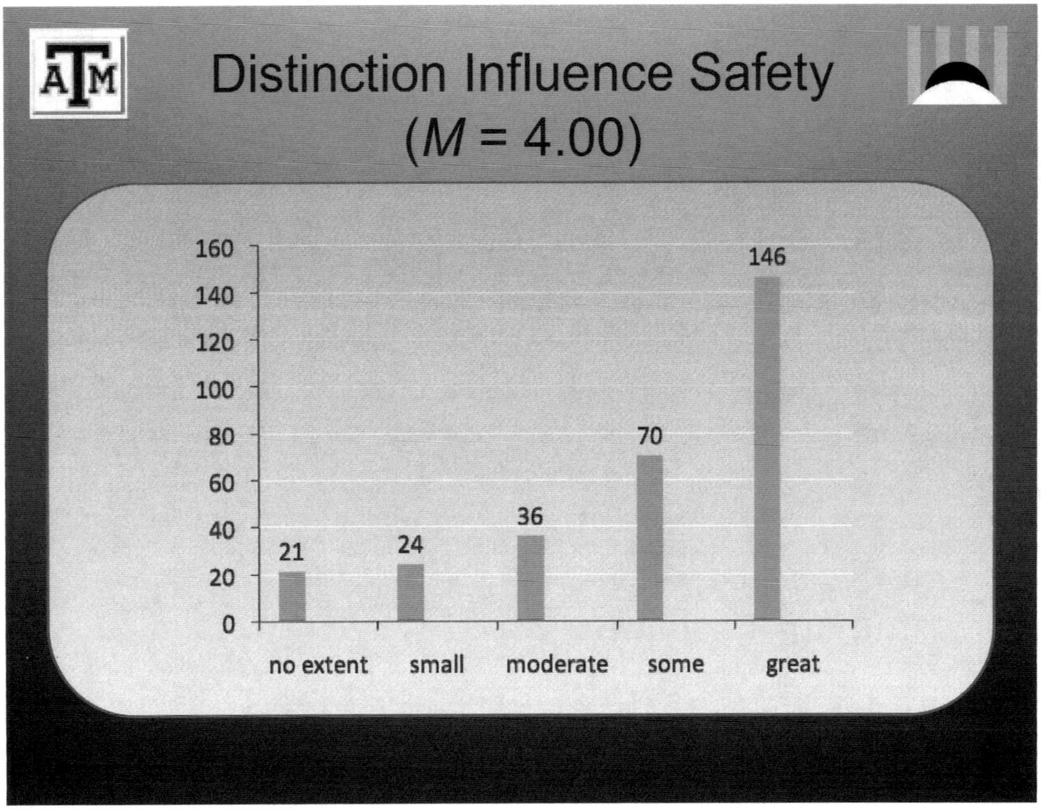
Distinction Influence Safety
(M = 4.00)
160
140
120
100
80
60
40
20
0
21
24
36
70
146
no extent
small
moderate
some
great

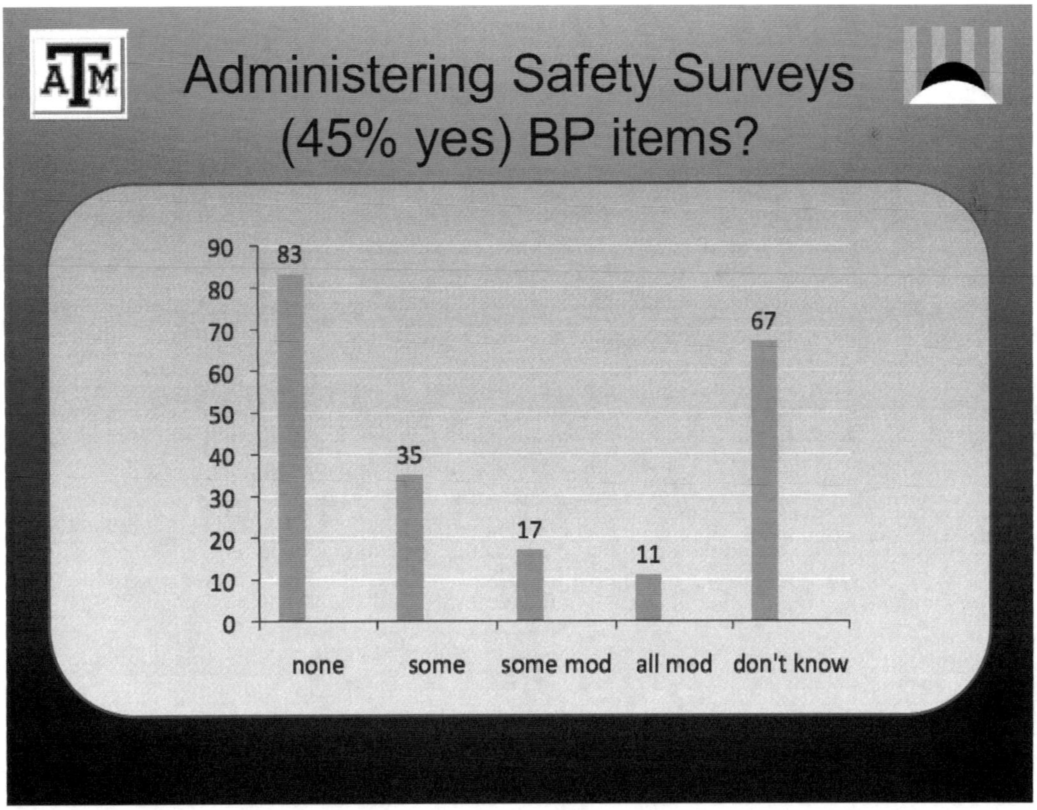
Administering Safety Surveys
(45% yes) BP items?
90
80
70
60
50
40
30
20
10
0
83
35
17
11
67
none
some
some mod
all mod
don't know

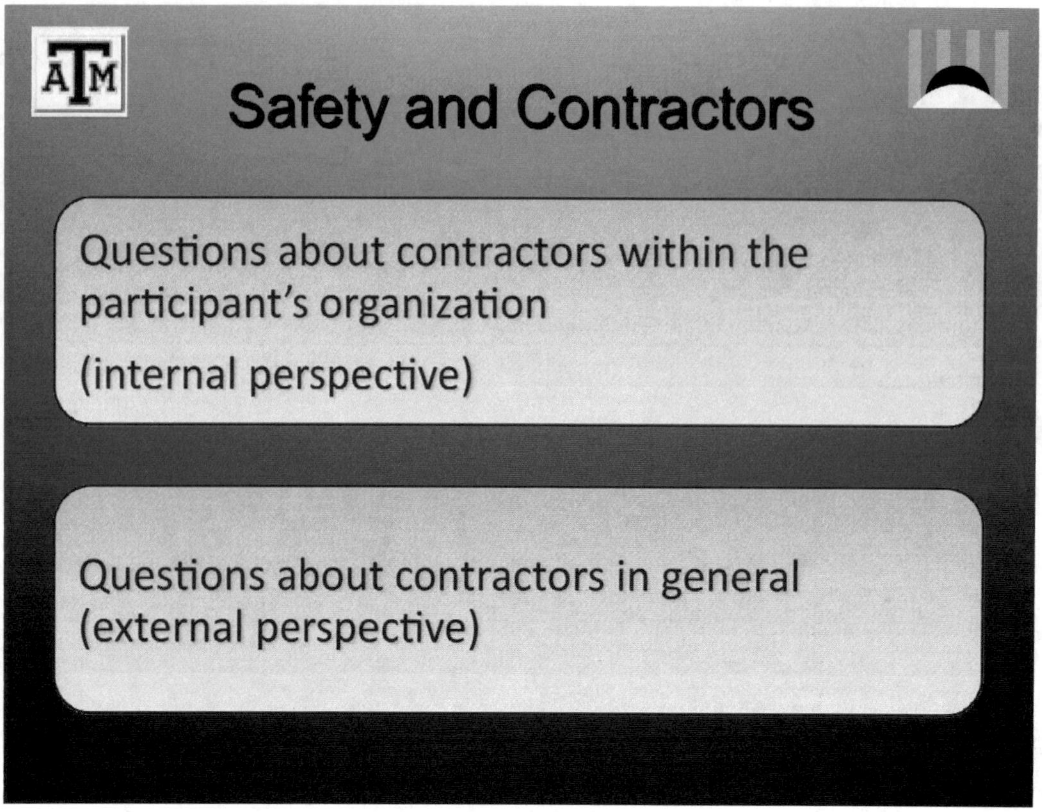
Safety and Contractors
Questions about contractors within the participant's organization
(internal perspective)
Questions about contractors in general
(external perspective)

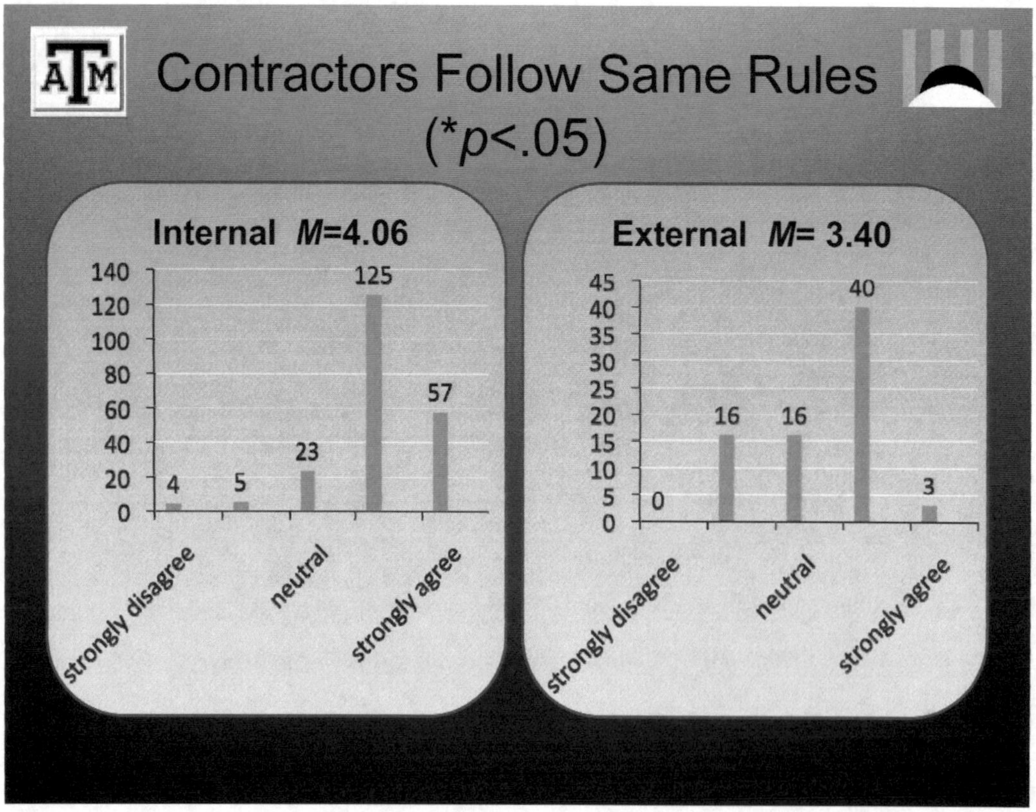
Contractors Follow Same Rules
(*p<.05)
Internal M=4.06
140
120
100
80
60
40
20
0
4
5
23
125
57
strongly disagree
neutral
strongly agree
External M= 3.40
45
40
35
30
25
20
15
10
5
0
0
16
16
40
3
strongly disagree
neutral
strongly agree

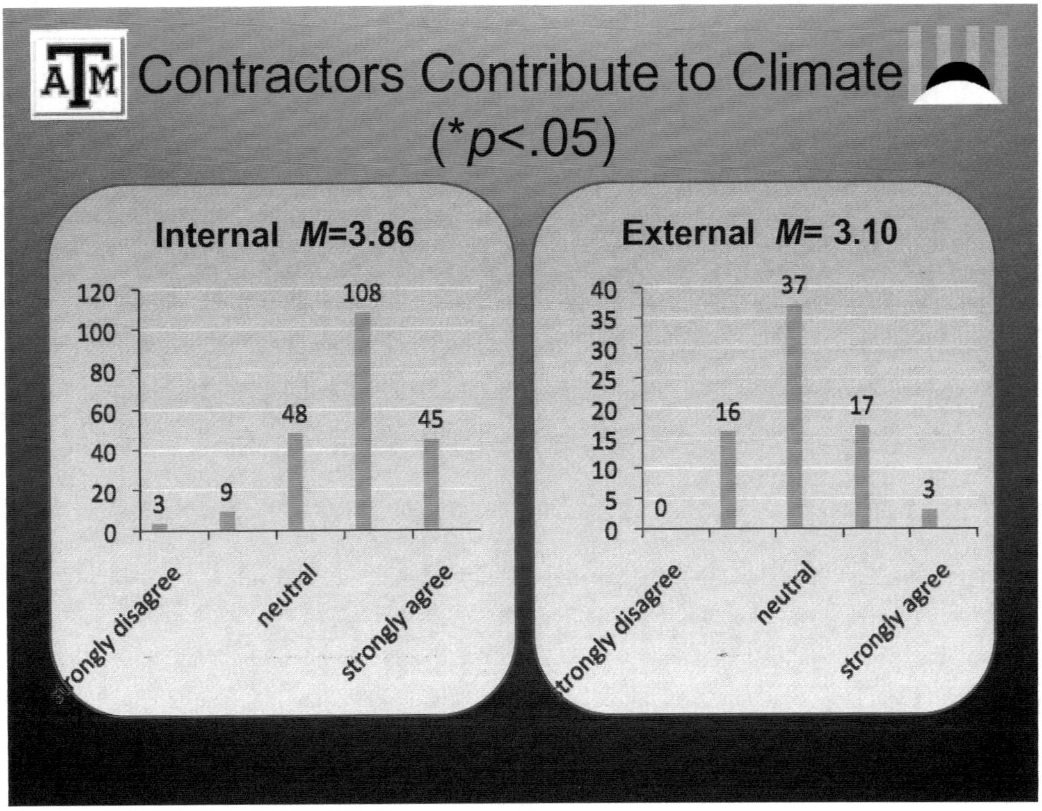

Contractors Contribute to Climate
(*p<.05)
Internal M=3.86
120
100
80
60
40
20
0
3
9
48
108
45
strongly disagree
neutral
strongly agree
External M= 3.10
40
35
30
25
20
15
10
5
0
0
16
37
17
3
strongly disagree
neutral
strongly agree

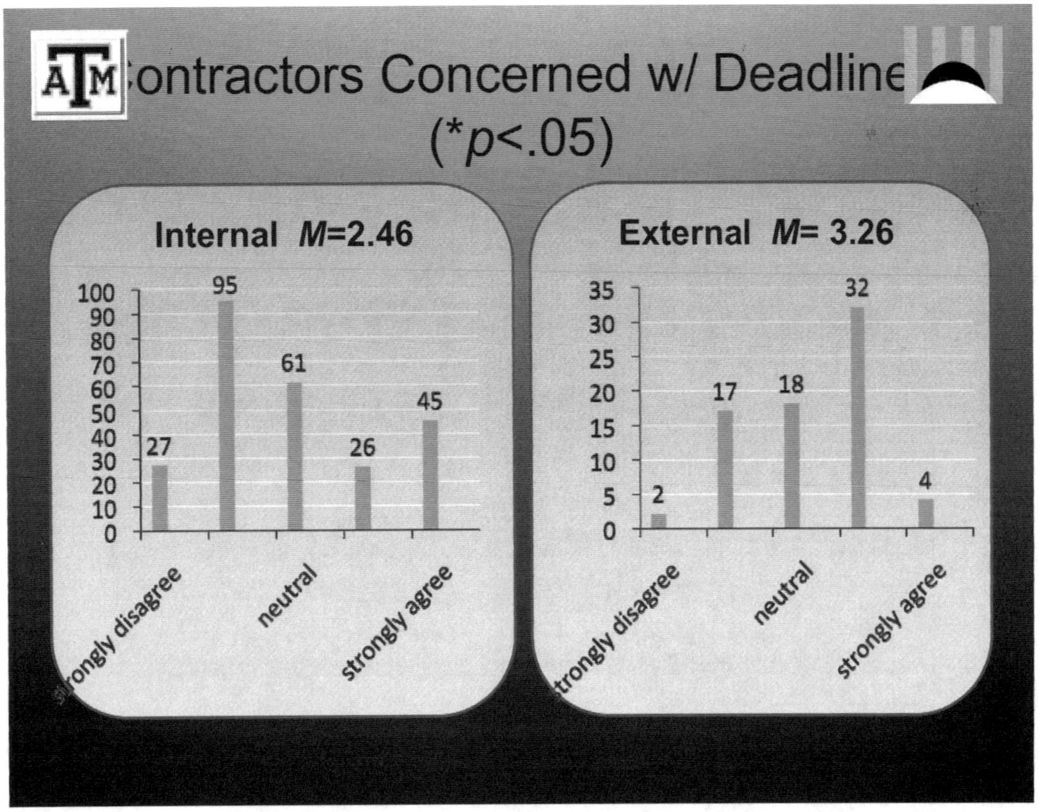

Contractors Concerned w/ Deadline
(*p<.05)
Internal M=2.46
100
90
80
70
60
50
40
30
20
10
0
27
95
61
26
45
strongly disagree
neutral
strongly agree
External M= 3.26
35
30
25
20
15
10
5
0
2
17
18
32
4
strongly disagree
neutral
strongly agree

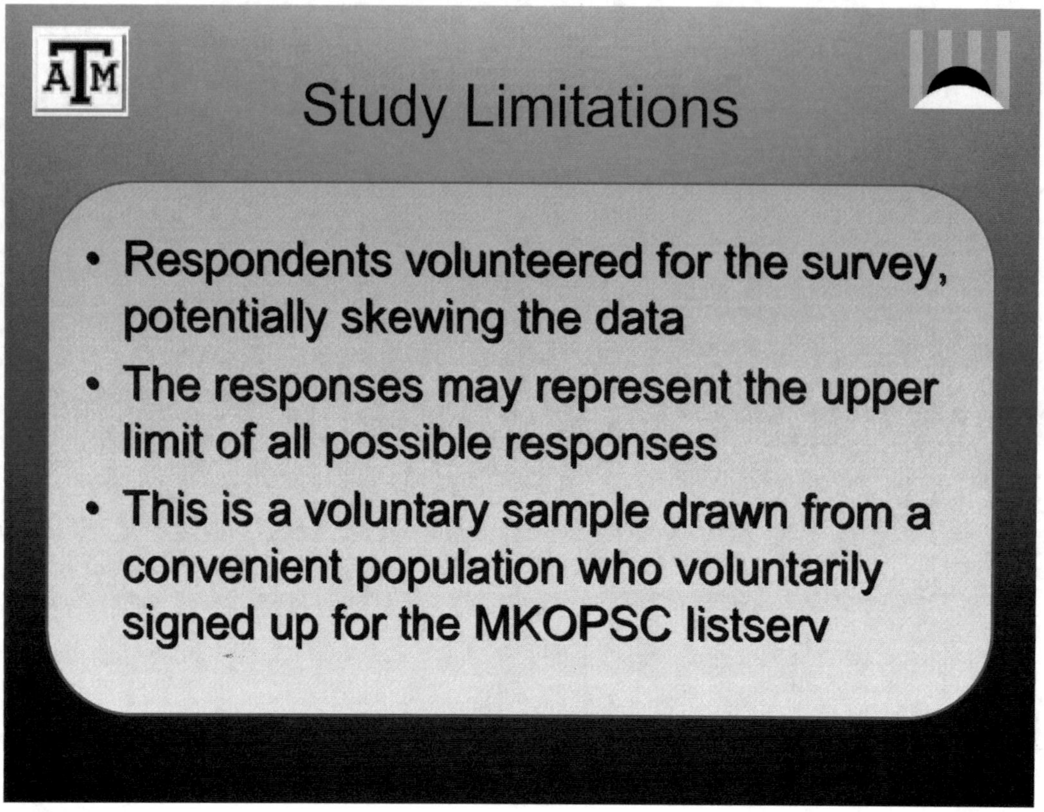
Study Limitations
• Respondents volunteered for the survey, potentially skewing the data
• The responses may represent the upper limit of all possible responses
• This is a voluntary sample drawn from a convenient population who voluntarily signed up for the MKOPSC listserv

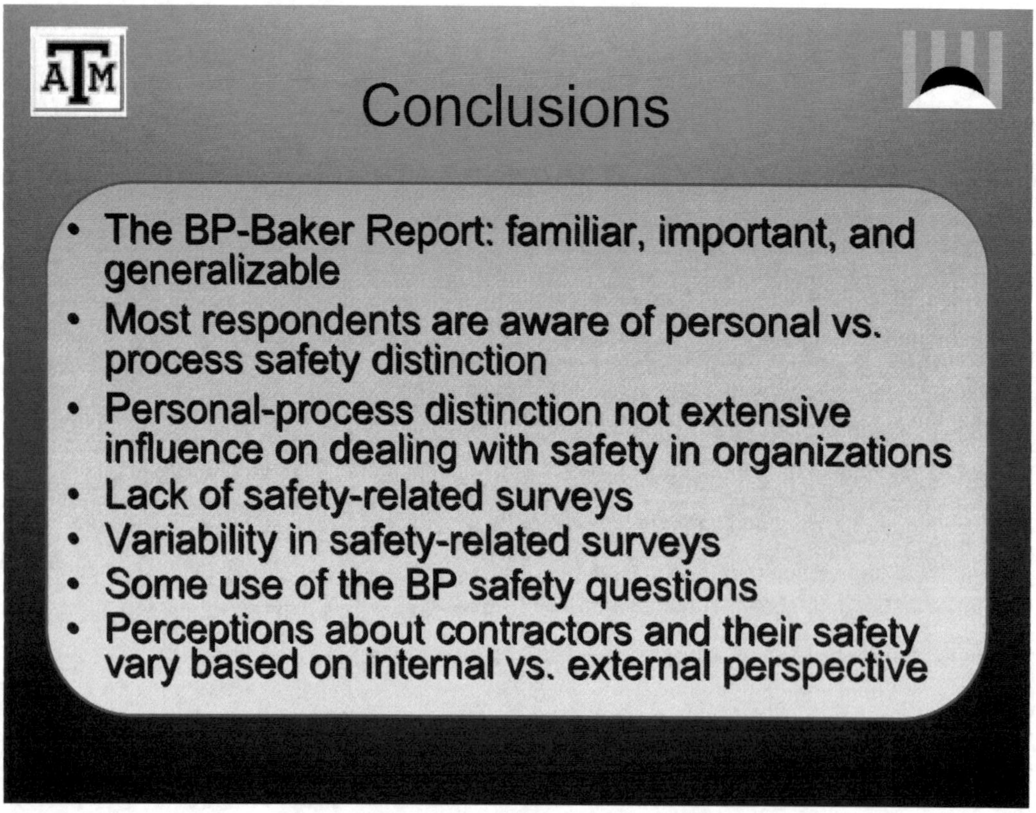
Conclusions
• The BP-Baker Report: familiar, important, and generalizable
• Most respondents are aware of personal vs. process safety distinction
• Personal-process distinction not extensive influence on dealing with safety in organizations
• Lack of safety-related surveys
• Variability in safety-related surveys
• Some use of the BP safety questions
• Perceptions about contractors and their safety vary based on internal vs. external perspective

Future Directions

- Examine physical proximity and affiliation with BP-TX City
- Code lessons learned and open-ended comments
- Continue to monitor the use of BP process safety culture survey questions
- Pursue explanations for the discrepancies in views about contractors between internal and external individuals

Process Safety Onshore and Offshore

Ian Sutton / Pat Campbell

AMEC Paragon

AIChE October 2009

Ian Sutton

- Engineering Safety Manager at AMEC Paragon in Houston
- Over thirty years of experience in design, start-up and operations of process plants
- Experience in offshore oil and gas, petrochemical plants, refineries, mining and alternative fuels
- [redacted] process risk management
- [redacted]ed, will publish next "Process Risk and Reliability Management"
- BS in Chemical Engineering for Univ. of Nottingham and MA in Literature from UHCL
- Registered Professional Engineer in Texas

Process Safety Onshore and Offshore

Ian Sutton / Pat Campbell
AMEC Paragon
AIChE October 2009

AMEC Paragon

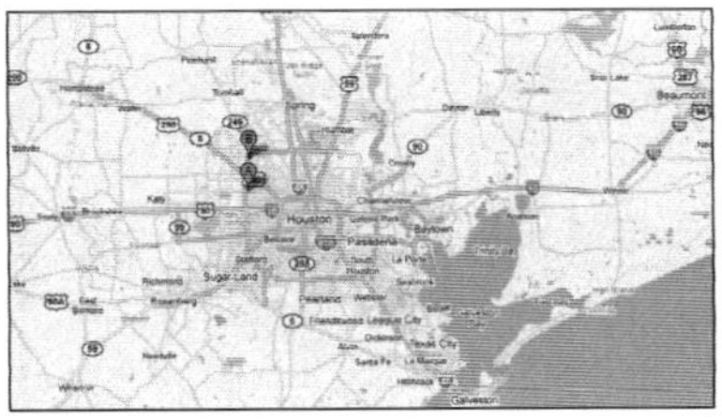

- Business Areas
  - Upstream
  - Alternative Energy
  - Pipelines
- AMEC
- Project Risk Management
  - Reliability, Availability, Maintainability (RAM)
  - Safety
  - Process Safety
  - Environmental

4

Purpose of Presentation

- Many similarities between onshore and offshore operations
  - Large volumes of hydrocarbons/chemicals, often at high pressure and temperature
  - Similar technologies
  - Potential for catastrophe / high public profile
  - Large investments
- For twenty years differing paths to process safety
  - Process plants onshore
  - Topsides (Gulf of Mexico)
  - Topsides (North Sea)
- Appears to be increased convergence
- Share lessons and insights
- Impact of SEMS

Structure of this Presentation

Daily Record
167 DEAD
DISASTER ON PIPER ALPHA
I FRIED IN THIS HELL

- Trends
- Features of offshore facilities
- Industry-changing events
- The regulatory environment
  - Onshore
  - Offshore
    - The Safety Case Regime
    - The API 14 "family"
    - MMS SEMS
- Offshore design and engineering

Trends: Onshore

Dramatic Improvement in Occupational Safety

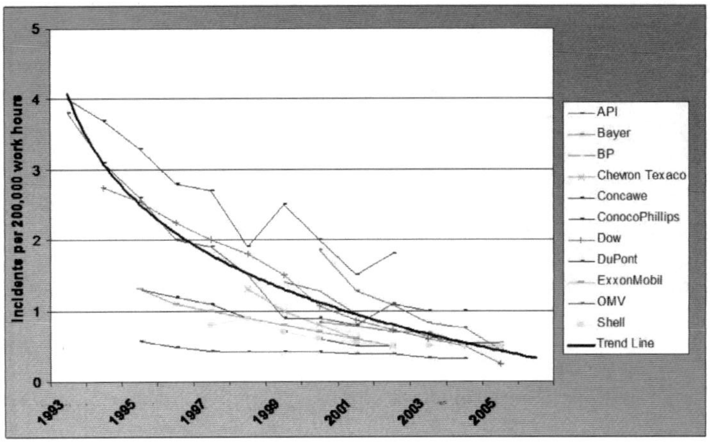

Courtesy DNV - 2008

7

Trends: Offshore

Dramatic Improvement in Occupational Safety

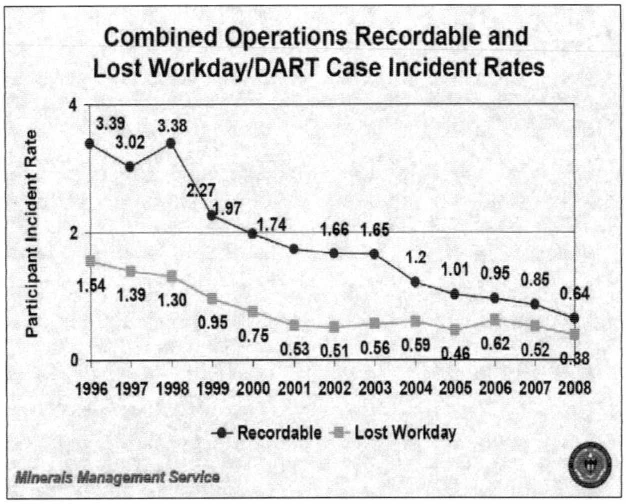

8

Features of Offshore Facilities

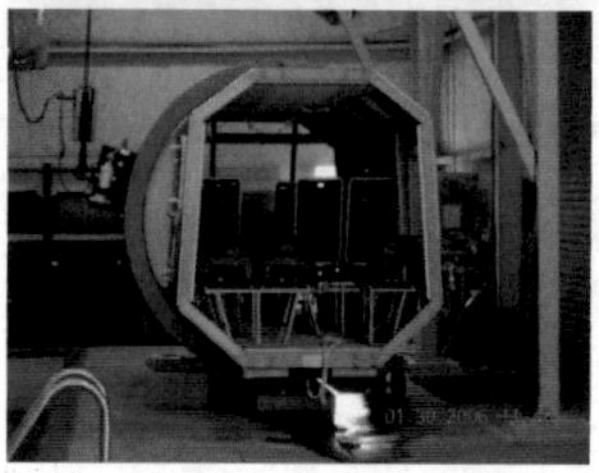

- Simple process steps
  - Separate oil, water, gas and sand
  - Possibility of H_2S
  - Some chemicals: glycol, methanol
- Generally low inventories (but FPSOs becoming more common)
- Congested
- Weight control critical
- Cannot easily escape
  - Lifeboats / rafts
  - Helicopters
  - The sea is usually cold
- Off-duty personnel are present

9

Gulf of Mexico (GoM) and Persian Gulf

- Started in shallow water, marsh platforms
- Benign environment
- Oil patch culture
- Now, very deep water (huge investment)
- Move toward safety cases

10

GoM: Lessons from Hurricanes

- Very costly
- Lose many platforms each season
- BUT: no fatalities, no injuries because we evacuate
- "If a man's not there he can't be killed".
- Downers and leaners very expensive to remediate

11

North Sea

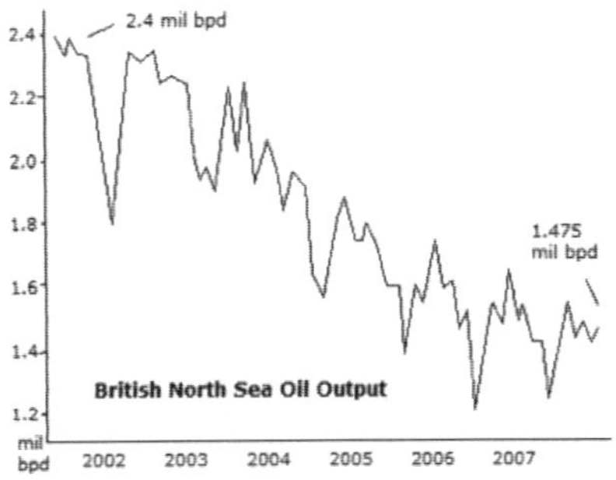

- Immediately in deep water
- Harsh environment
- No existing oil patch culture
- Production decline: smaller operators
- Move toward RP 14 approach

12

Industry Events

Onshore		Offshore	
1974	Flixborough	1984	API RP 14C
1976	Seveso (Safety Case)		
1984	Bhopal	1988	Piper Alpha
1992	OSHA PSM	1991	Cullen Report / Safety Cases
		1993	API RP 75
2005	Texas City		

The Regulatory Environment for Process Safety

- Both industries have pursued a non-prescriptive, performance-based approach
- Onshore: Regulations
  - OSHA 29 CFR 1910.119
  - EPA RMP
  - State regulations
- GoM: standards but no PSM rule (until now)
  - API RP 75: Recommended Practices for Development of a Safety and Environmental Management Program for Offshore Operations and Facilities (SEMP)
  - RP 14C
  - RP 14J
- MMS SEMS
- North Sea: The Safety Case Regime

RP 75 vs. OSHA's PSM

1. Safety and environmental information
2. Hazards analysis
3. Management of change
4. Operating procedures
5. Safe work practices
6. Training
7. Assurance of quality and mechanical integrity of critical equipment
8. Pre-startup review
9. Emergency response and control
10. Investigation of incidents
11. Audit of safety and environmental management program elements
12. Documentation and record keeping

1. Employee Participation
2. Process Safety Information
3. Process Hazards Analysis
4. Operating Procedures
5. Training
6. Contractors
7. Prestartup Safety Review
8. Mechanical Integrity
9. Hot Work
10. Management of Change
11. Incident Investigation
12. Emergency Planning and Response
13. Compliance Audits
14. Trade Secrets

15

MMS SEMS

- RP 75 creates a Safety and Environmental Management Program (SEMP): Guidance
- June 2009 MMS proposed a Safety and Environmental Management System (SEMS): Rule
  - Operating procedures
  - Mechanical integrity
  - Hazards Analysis
  - Management of Change

16

What Do HURRICANES and New Rules Have in Common?

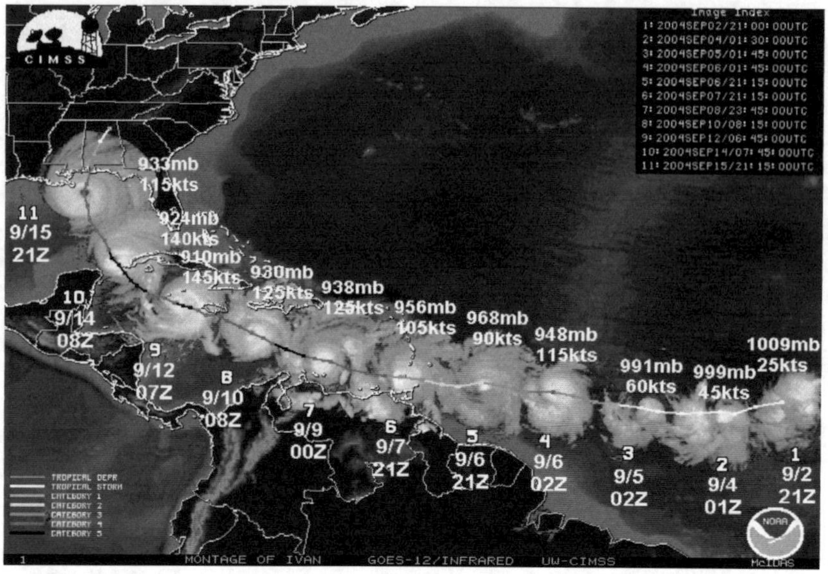

Offshore Operators Committee 2009

17

Both are Disruptive to Operations And are costly to Recover From!

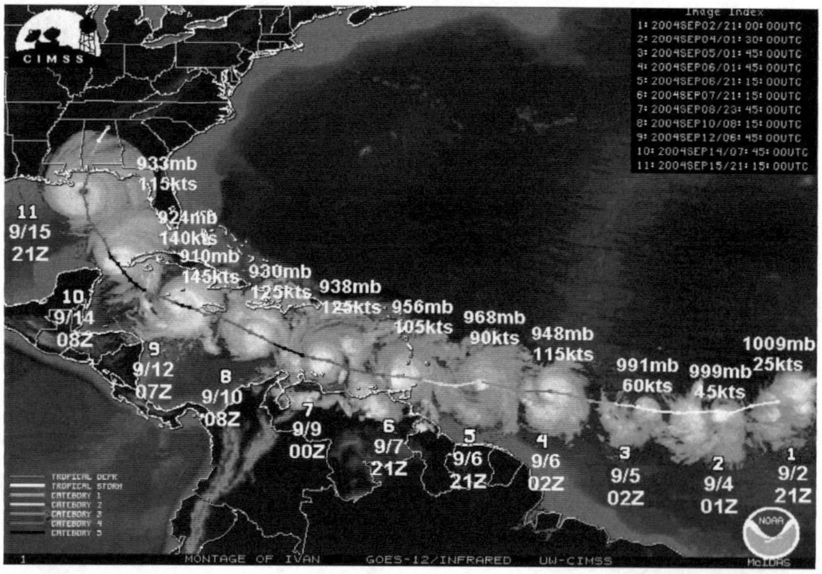

Offshore Operators Committee 2009

18

The Safety Case Regime

- "A documented body of evidence that provides a demonstrable and valid argument that a system is adequately safe for a given application and environment over its lifetime."
- Prepare three "cases" to demonstrate safety.
  1. Design
  2. Operations
  3. Shutdown
- After acceptance, implement a safety case "regime".
- Onshore: Seveso generated safety cases

19

Elements of a Safety Case

- Duty-Holder Responsibility;
- Auditor/Assessor Responsibility;
- Risk Management System;
- Management Systems; and
- Living Document

20

Design For Offshore Safety
US Gulf of Mexico (Federal Waters)

21

Authority

- In US Waters, the Federal Government has authority to lease and regulate extraction and production of natural resources on behalf of the United States.
- This is exercised by the Mineral Management Service (MMS) on behalf of the Department of the Interior (DOI).
- The governing document for this is the Code of Federal Regulations – 30 CFR 250.

22

Onshore and Offshore Similarities and Differences

Similarities

- OSHA requirements are enforced.
- Operating facilities have a duty to protect the public and their workers as well as the environment.

Differences

- Enforcement agencies are the MMS and the USCG rather than OSHA and EPA.
- Safety System requirements are contained in a limited number of industry documents and they are documented in the CFR.

- Offshore facilities are remote from populated areas.
- In the event of an incident, the offshore crew must rely on their own resources.

23

The API Family of Offshore Practices (The 14 Series)

RP 14B	■ Design, Installation and Operation of Subsurface Safety Valve System
RP 14C	■ Recommended Practice for Analysis, Design, Installation, and Testing of Basic Surface Safety Systems for Offshore Production Platforms
RP 14E	■ Design and Installation of Offshore Production Platform Piping Systems
RP 14F	■ Design and Installation of Electrical Systems for Fixed and Floating Offshore Petroleum Facilities for Unclassified and Class 1, Division 1 and Division 2 Locations
RP 14G	■ Fire Prevention and Control on Open Type Offshore Production Platforms
RP 14H	■ Use of Subsurface Safety Valves and Underwater Safety Valves Offshore
RP 14J	■ Design and Hazards Analysis for Offshore Production Facilities
RP 75	■ Development of a Safety and Environmental Management Program for Offshore Operations and Facilities

24

Purpose of Platform Safety System

The purpose of a production platform surface safety system is to protect:

- personnel,
- the environment,
- and the facility

from threats to safety caused by the production process.

The purpose of a safety analysis is to identify undesirable events that might pose a threat to safety, and define reliable protective measures that will prevent such events or minimize their effects if they occur.

25

Premises for Safety System Design (From RP 14C)

a. The process facility will be designed for safe operation in accordance with good engineering practices.

b. The safety system should provide two levels of protection to prevent or minimize the effects of an equipment failure within the process. *The two levels of protection should be independent of and in addition to the control devices used in normal process operation.* In general, the two levels should be provided by functionally different types of safety devices for a wider spectrum of coverage. Two identical devices would have the same characteristics and might have the same inherent weaknesses.

26

Premises (Continued)

c. The two levels of protection should be the highest order (primary) and next highest order (secondary) available. Judgment is required to determine these two highest orders for a given situation. As an example, two levels of protection from a rupture due to overpressure might be provided by a PSH and a PSL. The PSH prevents the rupture by shutting in affected equipment before pressure becomes excessive, and the PSL shuts in affected equipment after the rupture occurs.

However, *typically* a PSV is selected in lieu of the PSL because it prevents the rupture by relieving excess volumes to a safe location. Moreover, its fast response could prevent a rupture in situations where the PSH might not effect corrective action fast enough.

27

Underlying Logic

- If hydrocarbons are contained within the process, they pose no hazard as long as they are contained. Conditions that can breach the containment have the potential for escalation unless the energy release is controlled. The entire purpose of a process safety system is to first prevent, then contain any releases.

28

Modes Of Safety System Operation

- Automatic monitoring and automatic protective action if an abnormal condition indicating an undesirable event can be detected by a sensor,
- Automatic protective action if manually actuated by personnel who observe or are alerted to an unsafe condition by an alarm,
- And continuous protection by support systems that minimize the effects of escaping hydrocarbons.

- When an abnormal condition is detected in a process component by a safety device or by personnel, all input sources of process fluids, heat, and fuel should be shut off or diverted to other components where they can be safely handled.

29

Design Process

The design process goes through an evolution either formally or informally from the discovery/characterization of an offshore reservoir to its production.

The steps are:

- Evaluation
- Selection
- Definition
- Detail Design
- Fabrication
- Installation
- Commission
- Start-up/Production

30

ABS Consulting
AN ABS GROUP COMPANY
Safety, Risk and Integrity Management

Preparing for the OSHA Chemical NEP through RAGAGEP Compliance

October 1, 2009
Chad Patschke, Senior Risk/Reliability Engineer

Chad Patschke

- Sr. risk/reliability with ABS Consulting; previously with ICI Paints, ExxonMobil and Gulf Nuclear Industries
- Has over 15 years of experience in maintenance, mechanical integrity, PSM and risk analysis
- Assists clients with all aspects of Mechanical Integrity as well as development, implementation and auditing of PSM programs
- Has held positions of plant manager, operations manager, production superintendent, engineering and materials engineer
- BS in Engineering from Texas A&M and MS from George Washington Univ.

ABS Consulting
AN ABS GROUP COMPANY

ABS Consulting
AN ABS GROUP COMPANY

Safety, Risk and Integrity Management

Preparing for the OSHA Chemical NEP through RAGAGEP Compliance

October 1, 2009

Chad Patschke, Senior Risk/Reliability Engineer

3

Chemical National Emphasis Program

Directive 09-06 (CPL 02), effective July 27, 2009

ABS Consulting
AN ABS GROUP COMPANY

NEP Refining and Chemicals Comparison

DURATION	REFINING	CHEMICAL
Program	2 yr originally, + 2 year Region 6 + 8 months Region 8	1 yr. Pilot in Regions 1, 7, 10
		Likely to be nationwide after pilot (for ~7 yrs)
Site Visits	4 to 6 week on-site	1 week on-site
Inspection	4 to 6 months	1 to 3 months

ABS Consulting
AN ABS GROUP COMPANY

5

Lessons Learned from Refinery NEP

Directive CPL 03-00-010, reissued August 18, 2009

ABS Consulting
AN ABS GROUP COMPANY

Refinery NEPs

Citations have been primarily focused on the following areas:

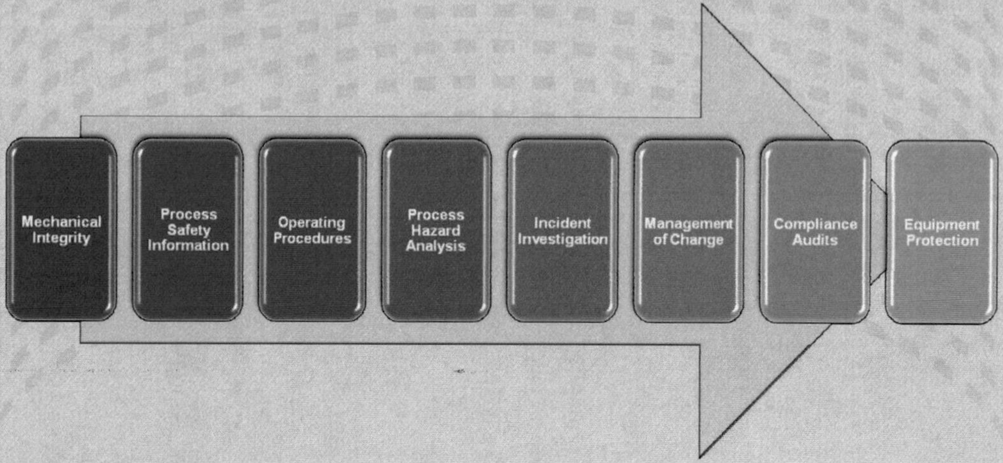

ABS Consulting
AN ABS GROUP COMPANY

7

NEP Citation Instances by Element

Element	Description	Number of Instances
J	**Mechanical Integrity (MI)**	**404**
D	**Process Safety Information (PSI)**	**270**
F	Operating Procedures (OP)	224
E	**Process Hazard Analysis (PHA)**	**195**
M	Incident Investigation (II)	159
L	Management of Change (MOC)	94
O	Compliance Audits (CA)	78
H	Contractors	35
C	Employee Participation	16
G	Training	15
	Total	**1489**

NEP Citation Instances by Sub-element

Sub Element	Description70	Number of Instances
d(3)	PSI – information pertaining to equipment in the process	246
j(4)	MI – Inspection and testing	198
e(3)	PHA – requirements of PHA	101
j(5)	MI – equipment deficiencies	95
j(2)	MI – written procedures	77

Referenced RAGAGEPs

- API
  - 510 PV inspection
  - RP 572 PV inspection
  - 521 PR & DP Systems
  - RP 520 PRD size, select, install
  - RP 576 Inspecting PRDs
  - 570 Piping inspection
  - RP 574 Piping inspection

ABS Consulting
AN ABS GROUP COMPANY

10

Referenced RAGAGEPs (cont.)

- API (cont.)
  - RP 578 PMI
  - RP 752 Building siting
  - RP 753 Temporary building siting
  - RP 579 FFS
- ASME
  - BPVC Sec VIII
  - BPVC Sec IX
  - B31.3

ABS Consulting
AN ABS GROUP COMPANY

11

Referenced RAGAGEPs (cont.)

- ANSI/ISA
  - S84.01 SIS (ITPM)
  - S91.01 ID'ing ESDs
- NFPA
  - 25 IT&M of Water Based Fire Protection Systems
  - 70 NEC
  - 101 Life Safety
  - 496 Purged & Pressurized Equip

ABS Consulting
AN ABS GROUP COMPANY

12

Referenced RAGAGEPs (cont.)

- CCPS
  - GLs for MI Systems
  - GLs for Hazard Evaluation
  - GLs for Engineering Design for Process Safety
  - GLs for Safe Process Ops & Maint
  - GLs for Safe Automation

ABS Consulting
AN ABS GROUP COMPANY

13

Referenced RAGAGEPs (cont.)

- CCPS (cont.)
  - GLs for Facility Siting & Layout
  - GLs for Evaluating Process Plant Buildings for External Fires & Explosions
  - Safe Design & Operation of Process Vents and Emission Control
  - LOPA

ABS Consulting
AN ABS GROUP COMPANY

14

Referenced RAGAGEPs (cont.)

- CSB
  - BP Investigation
  - Fire @ Formosa Case Study
  - Safety Bulletin - PMI
- Other
  - ASNT-TC1A
  - AIChe/Dow Fire & Explosion Index
  - Lee's Loss Prevention in the Process Industries

ABS Consulting
AN ABS GROUP COMPANY

15

(d)(3) Citation Examples

- P&ID's not accurate
- Relief system design & design basis not documented
- Failure to document that equipment follows RAGAGEP

ABS Consulting
AN ABS GROUP COMPANY

16

(d)(3) Citation Examples (cont.)

- Relief device set at 540 psig for 480 psig MAWP
- Relief isolation valve positions not controlled
- Relief valves undersized
- Deficiencies in IT&M of SIS

ABS Consulting
AN ABS GROUP COMPANY

17

(d)(3) Citation Examples (cont.)

- KO drum w/ inadequate retention time
- Trailers located subject to blast loading
- Positive pressure air supply to control building not alarmed
- Means of egress illumination not maintained

ABS Consulting
AN ABS GROUP COMPANY

18

j(4) Citation Examples

- All aspects of j(4) cited
- Many types of equipment & controls
- Many problems involved thickness measurements

ABS Consulting
AN ABS GROUP COMPANY

19

j(4) Citation Examples (cont.)

- Process piping not inspected per facility's I&T program
- I&T schedule on relief valves not adjusted when valves found to be heavily fouled
- Cathodic protection system & process analyzers not tested per employers program requirements

ABS Consulting
AN ABS GROUP COMPANY

20

j(4) Citation Examples (cont.)

- Anomalous thickness measurements not resolved per RAGAGEP (e.g., API 570)
- No inspection data for critical process piping circuit
- Thickness readings not documented

ABS Consulting
AN ABS GROUP COMPANY

e(3) Citation Examples

- Facility Siting
  - Failure to address potential atmospheric blowdown impacts
  - Flammable liquid piperack 22 ft from operator shelter
- Human Factors
  - failure to consider HF
  - failure to evaluate operator / DCS alarm interface

ABS Consulting
AN ABS GROUP COMPANY

e(3) Citation Examples (cont.)

- Did not consider freeze protection of low point piping
- Failed to address the lack of structural fireproofing near process equipment

ABS Consulting
AN ABS GROUP COMPANY

23

j(5) Citation Examples

- Pipe or equipment thickness measurements below retirement thickness
- Pressurized electrical cabinets w/o functional alarms
- Fireproofing insulation on ROSOV conduit not replaced
- Toxic / LEL gas detectors in control room air supply not working

ABS Consulting
AN ABS GROUP COMPANY

24

j(2) Citation Examples

- Under-insulation corrosion; injection points; deadlegs
- Non-metallic linings in pressure vessels
- Adequate number and locations of TMLs
- Relief valves past inspection intervals (many)

ABS Consulting
AN ABS GROUP COMPANY

25

j(2) Citation Examples (cont.)

- Performing repairs on pressure vessels (welder quals & certs, QC, etc.)
- Resolving "growth"

ABS Consulting
AN ABS GROUP COMPANY

26

What I see...

27

Relief Systems

- Code or standard referenced
- ALL scenarios considered
- Design basis (WCS)
- Physical properties / state of materials
- Atmospheric discharge in PHAs
- Car-seal procedure / verification

ABS Consulting
AN ABS GROUP COMPANY

28

Relief Systems (cont.)

- Flare & blowdown sizing calculations / updates (M&EB)
- Flare inspection records
- Radiant heat study
- Initial pop test
- Inspection frequency
- Postponing inspections (next T/A)

ABS Consulting
AN ABS GROUP COMPANY

29

Pressure Vessels & Piping

- Pipe classification
- TMLs – where and how many
- CUI, injection points, deadlegs
- Inspector qualifications and involvement
- Welder qualifications
- Below t-min (structural)
- Growth

ABS Consulting
AN ABS GROUP COMPANY

30

Pressure Vessels & Piping (cont.)

- Baseline data – date, thickness, CR
- Testing wrong locations
- Testing equipment calibration
- Deficiency management - postponing repairs
- Vessel liners (bonded, non-metallic)

ABS Consulting
AN ABS GROUP COMPANY

31

Electrical & Instrumentation

- Code / standard referenced
- Invisible barriers at unit boundaries/roads
- Identification and testing of SIS
- Testing fire detection / protection equipment
- Pressurized enclosures (buildings & cabinets) - design & ITPM

ABS Consulting
AN ABS GROUP COMPANY

32

PHAs & Siting

- Temporary buildings in overpressure zone
- Not completing HF & Siting checklists
  - Yes, No, blank
- Occupancy screening (man-hours)
- Building screening based on 1 psig

ABS Consulting
AN ABS GROUP COMPANY

33

Time for Questions

34

ABS Consulting

AN ABS GROUP COMPANY

Safety, Risk and Integrity Management

Chad Patschke
Process Safety Competency Center
CPatschke@absconsulting.com
(281) 673-2968

35

Jack Chosnek

Knowledge*One*

Committing Management to Process Safety

AIChE Regional Conference
Galveston, Texas
October 1st, 2009

Jack Chosnek

Knowledge*One*

- Independent consultant in Process Safety Management and Knowledge Management
- Over 33 years of experience in the petrochemical industry (20 in PSM)
- Worked for Celanese in PSM, Technology Management, Process Engineering, Operations, Projects and R&D
- BS and MS from Technion—Israel Institute of Technology, and PhD from Univ. of Missouri-Rolla, all in Chemical Engineering; MBA from Texas A&M C.C.
- Registered professional engineer in Texas

Jack Chosnek
Knowledge*One*

Committing Management to Process Safety

AIChE Regional Conference
Galveston, Texas
October 1st, 2009

Committing Management to Process Safety

Topics

- The challenge
- Proposed solution
- Metrics and indicators
- Implementation

Knowledge*One*

Challenges to Management PSM Commitment

- Define "good" process safety management
- Make PSM a part of manager's performance

KnowledgeOne

Performance Indicators

- Easy to understand
- Similar to financial indicators
- Can be measured throughout the year

KnowledgeOne

Existing Indicators and Proposed Metrics

Existing Indicators

- Injury Indices
- Regulatory Compliance (OSHA 1910.119)
- Safety Culture (~)

Proposed Metrics

- CCPS (Leading and Lagging)

KnowledgeOne

Our Proposal

- Manage Process Safety on a Risk Basis
- Provide an Indicator of Total Risk

KnowledgeOne

Elements of Risk

- Action Items Follow-up
- Management of Change (MOC)
- Process Safety Training, and
- Operational Integrity

KnowledgeOne

Elements/Metrics

Action Items Follow-up

- Number of Open Action Items from
  - PHAs
  - Incident Investigations
- Incomplete Incident Investigations

KnowledgeOne

Elements/Metrics

Management of Change (MOC)

- Number of Open MOCs
- Number of Past-Due MOCs
  - Temporary
  - Regular

KnowledgeOne

Elements/Metrics

Process Safety Training

- Pct. of Operators Trained in Quarter
- Pct. of SOPs trained on in Quarter

KnowledgeOne

Elements/Metrics

Operational Integrity

- No. of "Critical Unavailabilities" (weekly)
  - Equipment
  - Instrumentation
  - Operators

KnowledgeOne

Conclusions

- An indicator provides management a tool for:
  - understanding the health of PSM
  - communicating it both to employees and top management
- The indicator helps to assess the manager's PSM performance
- Both factors push managers towards PSM commitment

KnowledgeOne

Proposed CCPS Metrics

- Maintenance of Mechanical Integrity
- Action Items Follow-up
- Management of Change (MOC)
- Process Safety Training and Competency
- Safety Culture

KnowledgeOne

Correlation of CCPS Metrics

- Correlate with Lagging Indicators
  - Near Misses
  - Actual Incidents

KnowledgeOne

AIChE STS Galveston Conference

Advances in Petrochemical Technology

Olefin – Refinery Integration

Colin Bowen
(Dilip Dharia)
The Shaw Group

1 October 2009

Olefin – Refinery Integration

Key Issues

- Lower feedstock cost (VGO vs naphtha)
- Related ethylene, propylene costs vs Steam Cracking (SC)
- Recent FCC derived propylene, ethylene productions
- Increasing propylene demand / supply
- Future FCC versions including High Olefin FCC (HOFCC)
- Low cost Refinery Off Gas (ROG)
- Convenient co-product recycle cracking

FCC & SC Related Feed/Product Costs (USA)

		June 2008	Jan 2009	Sep 2009
Crude Oil	$/BBL	140	~40	~ 67
	$/MMbtu	25	~7	~12
	¢/lb	45	~13	~22
Ethane	¢/lb	55	20	~23
Light Naphtha	¢/lb	76	29	~29
Gas Oil	¢/lb	93	55	~23
HVGO	¢/lb	40	11	~14
ATB	¢/lb	~30	~9	~12
Ethylene	¢/lb	70	21	~37
Propylene (PG)	¢/lb	82	18	~57

CMAI

HVGO - Hydrotreated Vacuum Gas Oil
ATB - Atmospheric Tower Bottoms

3

Shaw® a world of Solutions™

The Shaw Group Inc.®

Global Ethylene Production & Predictions

2008 Production **~ 118 MMTA (Capacity ~ 127 MMTA)**

2012 Production **~ 140 MMTA (Capacity ~ 160 MMTA)**

2012 % Production from Ethane ~ 43%

- **Mid East Project Feedstock Incentives**
- **US Ethane vs Naphtha Cost Advantage**

Current USGC Price

July 09 ~ 37¢/lb

Sept 09 ~ 37¢/lb

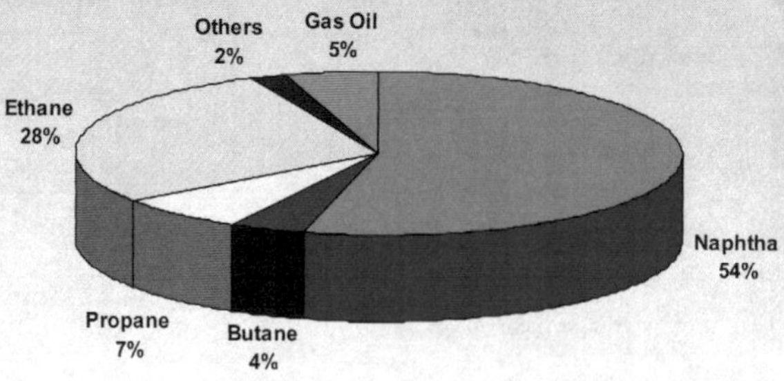

CMAI

4

Shaw® a world of Solutions™

The Shaw Group Inc.®

Global Propylene Production & Predictions

2008 Production	~ 72 MMTA
2012 Production	~ 92 MMTA
Δ Production from Pyrolysis	~ 7 MMTA
Δ Production from FCC linked units	~11 MMTA
Δ Production from Alternative Routes	~ 2 MMTA

Current USGC Price

July 09 ~ 41 ¢/lb

Sept 09 ~ 57 ¢/lb

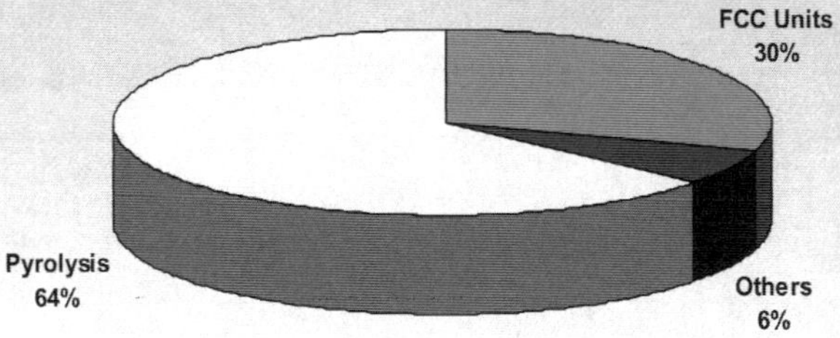

5

Shaw a world of Solutions

The Shaw Group Inc.

To Meet Increasing Demands of Propylene

- Attain higher P/ E ratio in the steam cracker
- Extend the steam cracker olefins feedstock flexibility
- Maximize production of olefins from refinery-based heavier feedstocks
- Recover ethylene and propylene from FCC off gases
- Recycle refinery co-product streams, eg ethane, propane, etc., to SC
- Convert C_4, C_5 mono-olefins to propylene

6

Shaw a world of Solutions

The Shaw Group Inc.

Global FCC Production

- Global FCC operating capacity ~ 16 MMBPD (~770 MMTA)
- USA capacity ~ 40%
- Propylene : Gasoline ratio can be increased by catalyst adjustment (ZSM-5)
- Diesel : Gasoline demand ratio is increasing
- Future FCC units will expand HOFCC capacity
- HOFCC : FCC ratio is increasing significantly

7

Shaw® a world of Solutions™

The Shaw Group Inc.®

Catalytic Cracking Processes –Key Elements

Objective is to produce ethylene, propylene and transportation fuels

- **Fluid Catalytic Cracking (FCC) – high propylene option**- transportation fuels plus some propylene from VGO type feedstocks
- **Residual Fluidized Catalytic Cracking (RFCC) - high propylene option**- transportation fuels plus some propylene from heavy feedstocks
- **Deep Catalytic Cracking (DCC, Sinopec RIPP)** propylene plus transportation fuels from high quality VGO type feedstocks
- **Catalytic Pyrolysis Process (CPP, Sinopec RIPP)** ethylene and propylene plus transportation fuels from good quality heavier feedstocks

Shaw® a world of Solutions™

The Shaw Group Inc.®

FCC Basic Technology Features

Converter
- Riser/Injection/Termination
- Spent Catalyst Stripper
- Regenerator(s)
- Flue Gas Scrubber

Recovery
- Main Fractionator
- Absorption System
- Propylene Treatment/Distillation
- Off Gas Treatment

9

Shaw a world of Solutions

The Shaw Group Inc.

Typical Yields (wt% from feed)

	Propylene	Ethylene	Butylene	Gasoline
Standard FCC Unit	4-6	~ 1	6-8	~ 55
HOFCC Unit	10-12	~ 2	7-10	~ 42
DCC-2 Unit	12-14	~ 3	11- 14	~ 41
DCC-1 Unit	18-24	~ 5	11-14	~ 27
CPP Unit	16 - 25	10 - 24	7 - 14	15 - 18

10

Shaw a world of Solutions

The Shaw Group Inc.

Latest DCC Capacities (1997 – 2011)

Unit	Feedstock Type	Feedstock (kta)	Feedstock (BPSD)	C_3^- (kta)	C_2^- (kta)
TPI	VGO/ATB	900	19,000	140	45
PetroRabigh	HVGO	4300	92,000	820	260
HMEL	HVGO	2200	47,000	430	110
Nishnekamsk	HVGO	1100	23,000	225	60
MRPL	HVGO	2200	45,000	430	110

11

Shaw® a world of Solutions™

The Shaw Group Inc.®

Latest HOFCC Capacities (2009 – 2013)

Unit Location	Feedstock Type	Feedstock BPSD	Feedstock KTA	Propylene (KTA)
Mid-East	VGO	95,000	4800	557
Mid-East	VGO	38,000	1800	180
Mid-East	Resid	127,000	6000	750
Mid-East	VGO	51,000	2700	250
Egypt	Resid	68,000	3200	320
Mid-East	Resid	60,000	2800	330
Mid-East	VGO	33,000	1550	160
Korea	Resid	75,000	3500	380
Taiwan	Resid	80,000	3800	480
Mid-East	Resid	91,000	4300	210
Vietnam	Resid	65,000	3050	150

12

Shaw® a world of Solutions™

The Shaw Group Inc.®

RFCC, HOFCC

Steam Cracker (SC) Liquid Feedstocks Recent Ethylene, Propylene Capacities

UNIT	Ethylene (kta)	Propylene (kta)
China	600	325
China	800	430
KSA, Mid- East	1300	450
China	450	266
China	800	400
China	800	400
Brazil	1040	507
Singapore	~1000	~640

Mega Ethylene Recent Units

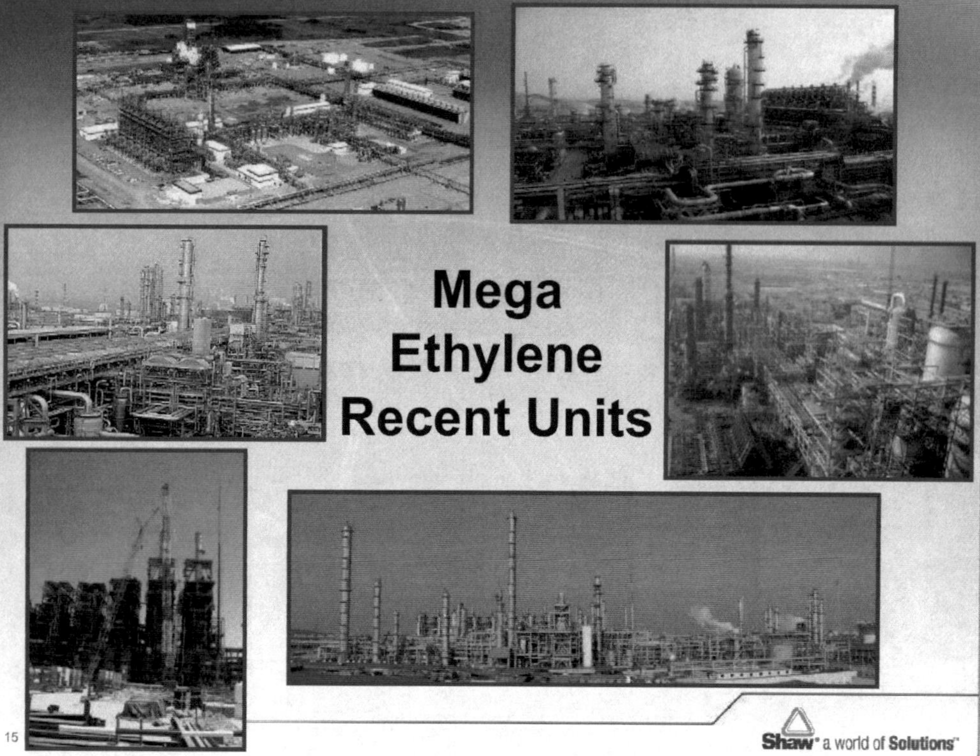

15

Shaw® a world of Solutions™

The Shaw Group Inc.®

DCC Technology Features

- Proven FCC elements from Shaw/Axens FCC process
- Secondary cracking of naphtha to produce olefins
- No separate additive / diluent
- Competitively priced high pentasil catalyst
- High yields of C_3^-, C_4^- olefins and LPG
- Extensive Shaw experience in polymer grade C_3^- and C_2^- production including contaminant removals
- Well-proven process; ~16 commercial units in operation/design
- Suitable for integrating with steam crackers

16

Shaw® a world of Solutions™

The Shaw Group Inc.®

DCC - Thai Petrochemical Industries (1997)

Feedrate	**18 MBPD**
Propylene	**150 KTA**
Feedstocks	**VGO**
Facilities	**HTU, ERU, PRU**

17 Shaw® a world of Solutions™

The Shaw Group Inc.®

DCC – PetroRabigh - KSA (2009)

- Integrated DCC & ethane SC
- Startup: Mid 2009
- Feedrate: 92,000 BPD
- Feedstocks: HVGO
- Reactor: 15 meter diameter
- Regenerator: 17 meter diameter
- Propylene: 950 kta (total)

18 Shaw® a world of Solutions™

The Shaw Group Inc.®

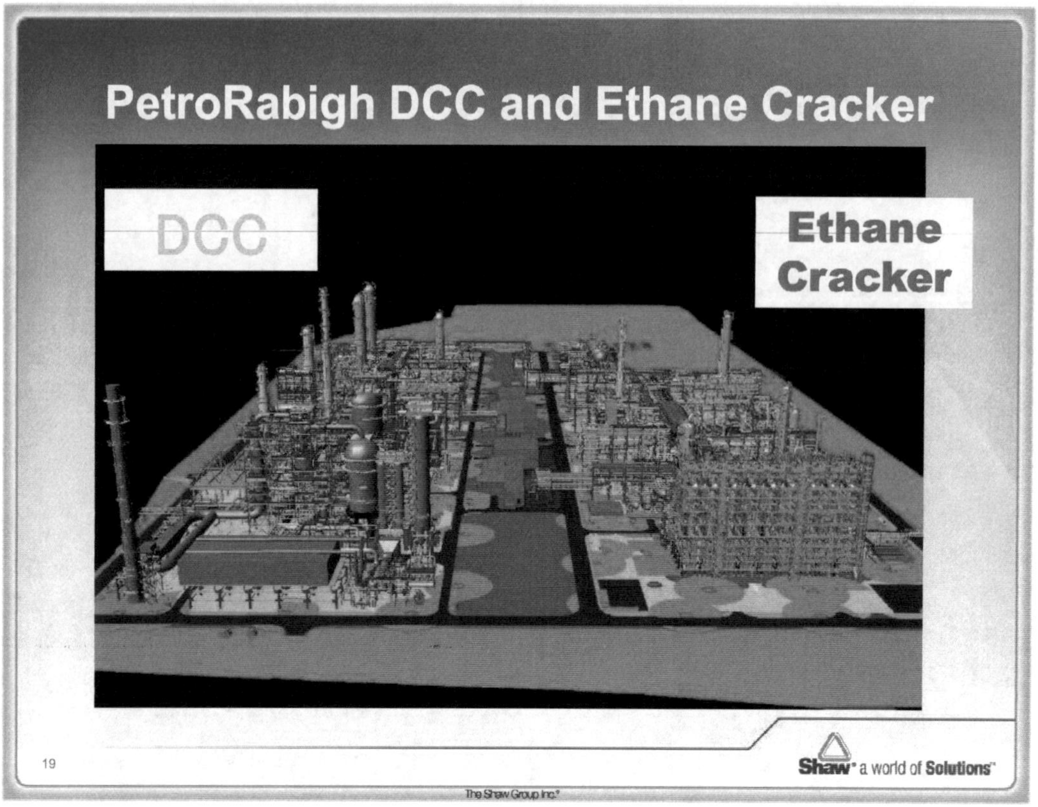
PetroRabigh DCC and Ethane Cracker
DCC
Ethane
Cracker
19
Shaw a world of Solutions
The Shaw Group Inc.

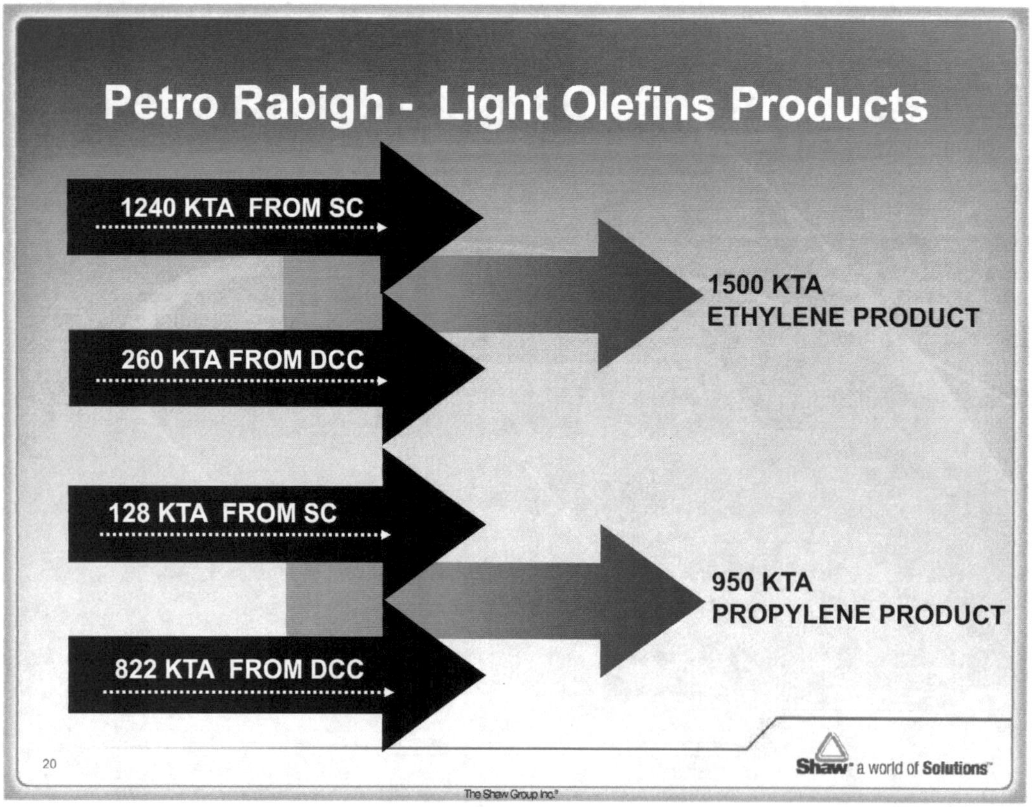
Petro Rabigh - Light Olefins Products
1240 KTA FROM SC
260 KTA FROM DCC
1500 KTA
ETHYLENE PRODUCT
128 KTA FROM SC
822 KTA FROM DCC
950 KTA
PROPYLENE PRODUCT
20
Shaw a world of Solutions
The Shaw Group Inc.

Product Yields for DCC and FCC Types

Material Balance wt% (Typical HVGO Feed)	DCC - 1	DCC - 2	FCC
C_2 Minus	11.2	3.6	3.5
C_3 & C_4 (LPG)	42.5	34.5	17.7
Naphtha	26.5	42.0	54.9
LCO (light cycle oil)	9.5	9.8	10.3
DO (decant oil)	4.3	5.8	9.3
Coke	6.0	4.3	4.3
Total	**100.0**	**100.00**	**100.00**
Ethylene	5.5	1.5	0.9
Propylene	19.5	14.3	4.9
Butylene	13.5	14.7	8.1
P/E	3.5	9.5	5.5

21

Shaw a world of Solutions

The Shaw Group Inc.

Update on DCC Catalyst

- **1st Generation DCC Catalyst**
  CHP: Propylene selectivity
- **2nd Generation DCC Catalyst**
  CRP: Hydrothermal stability
  CIP: Higher activity
  Bottoms cracking
  Metals tolerance
- **3rd Generation DCC Catalyst**
  MMC: New generation catalysts
- **4th Generation DCC Catalyst**
  DMMC: Heavy oil conversion and higher propylene yield

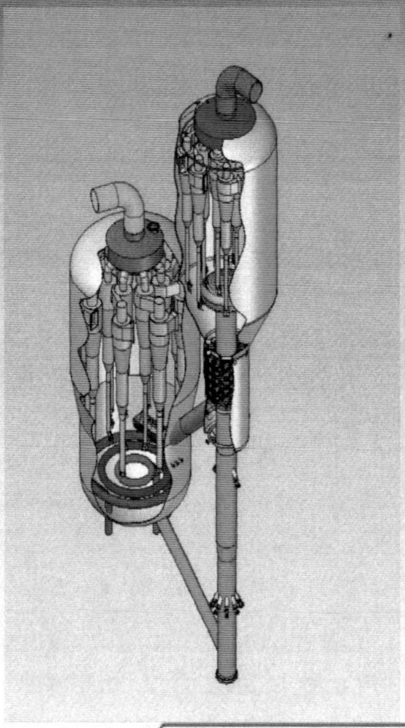

22

Shaw a world of Solutions

The Shaw Group Inc.

Summary of DCC Process Advantages

- DCC is still the only commercially proven process to produce high yields of propylene (18 to 24 wt%) from VGO.
- DCC provides optimum integration between the refinery and steam cracker ; very attractive return on capital.
- Low cost feedstocks and operational flexibility to technically and financially product fuels and petrochemicals.

23

Shaw® a world of Solutions™

The Shaw Group Inc.®

Catalytic Pyrolysis Process (CPP) Technology

- FCC type reactor / regenerator scheme
- Dual function type catalyst
  - Acid sites, carbenium ions, C_3^- C_4^- production
  - Free radical initiators, C_2^- C_3^- production
- Higher severity cracking of heavy oil
- Secondary cracking of naphtha
- Recycle cracking of low value co-products
- Linked olefin recovery, contaminant removal systems
- Flexible operating modes
  - Maximum ethylene
  - Maximum propylene
  - Maximum total olefins

24

Shaw® a world of Solutions™

The Shaw Group Inc.®

Catalytic Olefin Cracking Conditions

Catalytic Olefin Performance Parameters			
	FCC/RFCC	DCC	CPP
Reactor Temp (°C)	500-550	530 - 580	580-640
Reactor Pressure (barg)	1-2	0.7 – 1.5	0.8-1.0
Residence Time (sec)	1-3	2 - 8	2-8
Catalyst:Oil Ratio (wt)	5-10	8 - 15	15-20
Steam:Hydrocarbon (wt)	1-6	10-30	30-50
Riser and Reactor Bed	Riser	Riser & Bed	Riser & Bed

25

Shaw® a world of Solutions™

Catalytic Pyrolysis Process

Demonstration Unit Data (Quality ATB Feed)

Operating Conditions, Product Distribution, and Olefin Yield in wt%

Operating Mode	**Mode 1**	**Mode 2**	**Mode 3**
Reaction temp, °C	576	610	640
Catalyst / oil ratio	14.5	16.9	21.1
Steam / oil ratio	0.30	0.37	0.51
H_2S	0.08	0.08	0.08
C_2 minus	17.7	26.5	37.5
$C_3 + C_4$	44.0	36.9	28.7
C_5 + naphtha	17.9	17.8	15.0
LCO	11.8	9.1	8.0
Coke	8.5	9.8	10.8
	100.00	100.00	100.00
Ethylene	**9.8**	**13.71**	**20.37**
Propylene	**24.6**	**21.5**	**18.2**
Butylenes	**13.2**	**11.3**	**7.5**

Shaw® a world of Solutions™

CPP/DCC Integrated Ethylene Plant

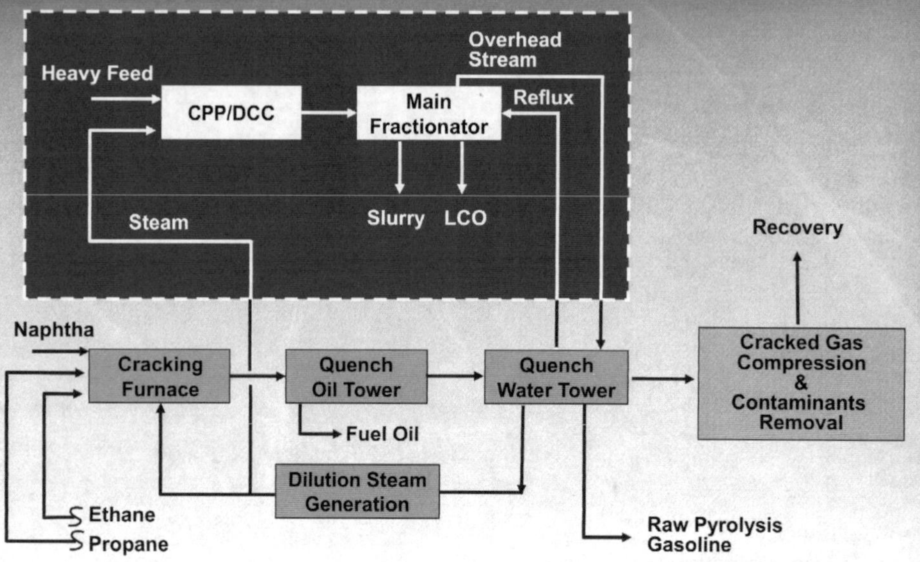

27

Shaw a world of Solutions

The Shaw Group Inc.

Shenyang Paraffin-Wax Chemical Co., Ltd CPP Commercial Prototype 2009

Earlier DCC-2 Unit; First Commercial CPP Unit

28

Shaw a world of Solutions

The Shaw Group Inc.

CPP Advantages & Potential

- Low cost feedstock (ATB, VGO, etc)
- High quality C_5^+ aromatic naphtha
- Flexible P:E ratio
- Recycle cracking of low value co-products
- Low cost fuel (i.e. self-produced coke)
- Integrate with existing ethylene unit

29 Shaw® a world of Solutions™

CPP Recovery Design Principles

- CG compression, 4 stages
- Amine / NaOH acid gas removal
- Front end DeC_3, As/Hg removal
- De-Oxo, drier, COS/RSH chemi-sorption
- Conventional cryogenic fractionation
- C_3 stripper / rectifier / hot water belt
- C_2, C_3, C_4, C_5 recycle cracking furnace

Optimally Combined
Steam Cracker / ROG Configuration

30 Shaw® a world of Solutions™

Integration Design Issues (1)

- Independent or linked units (SC-DCC-CPP)
- Absorbers vs Fractionators
- SC (Steam Cracker) contaminants
  - Acetylenes, Dienes
- DCC/CPP contaminants
  - O_2, NOx, RSH, NH_3, COS, arsine, CO_2, H_2S, metals, etc.
- Correct sequencing of treatment systems
- Optional recycle cracking
  - Ethane, propane, C_4's, C_5's

31

Shaw® a world of Solutions™

The Shaw Group Inc.®

Integration Design Issues (2)

- Linked quench oil / cycle flux oil
- Common dilution steam generation
- Linked wet gas / cracked gas compression
- Combined C_2^- C_3^- fractionation & refrigeration
- Combined light ends fractionation
- Combined contaminant removal
- Multi-recycle feed options
- Fuel system optimization

(Reference US Patent 5,523,502, Flexible Light Olefins Production)

32

Shaw® a world of Solutions™

The Shaw Group Inc.®

Refinery Off Gas Opportunities

- Refinery Off Gas (ROG) primary source is FCC
- FCC derived C_2H_4 is ~ 0.7 – 1.2% wt
- Hence potential ROG C_2H_4 is up to 7 MMTA (but actual rate is less than 1 MMTA)
- ROG collection and ethylene recovery is advantageous where refinery / petrochemical units are adjacent
- ROG recovery units in several such USGC sites

Refineries and Steam Crackers - USGC

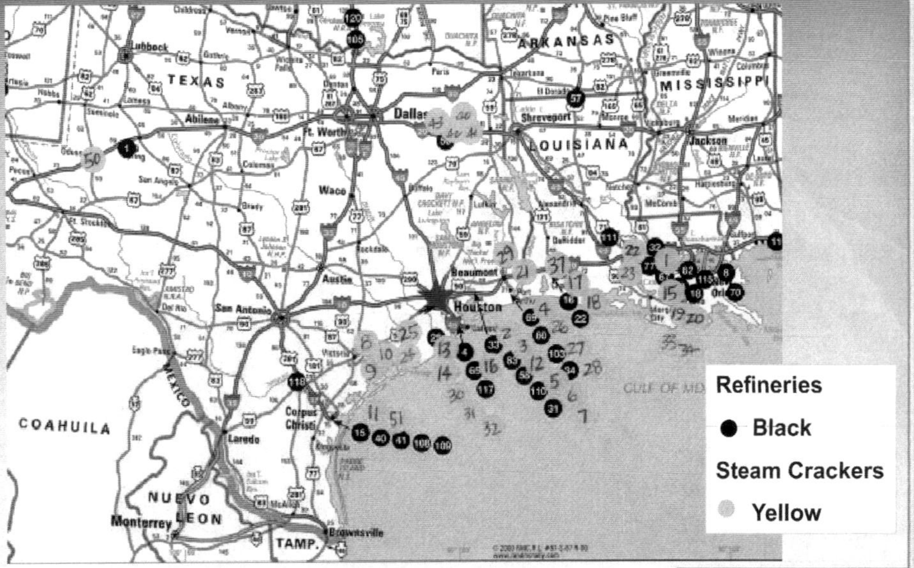

Typical Contaminants in ROG

O_2	Thiophenes
NOx	Disulfides
H_2S	MeOH
CO_2	Propanol
H_2O	Mercury
COS	AsH_3
CS_2	PH_3
NH_3	Oxygenates
HCN	Acetylenes
RSH	Dienes

Proprietary ROG Purification Designs for C_2 lighter & C_3,C_4

35

Shaw® a world of Solutions™

The Shaw Group Inc.®

Recent ROG Recovery Unit

38

Shaw® a world of Solutions™

The Shaw Group Inc.®

Recent ROG Recovery Unit
DeOxo Reactor
< 1 ppm mol O_2
< 0.1 ppm mol NOx
< 0.5% C_2H_4 loss
39
Shaw a world of Solutions
The Shaw Group Inc.

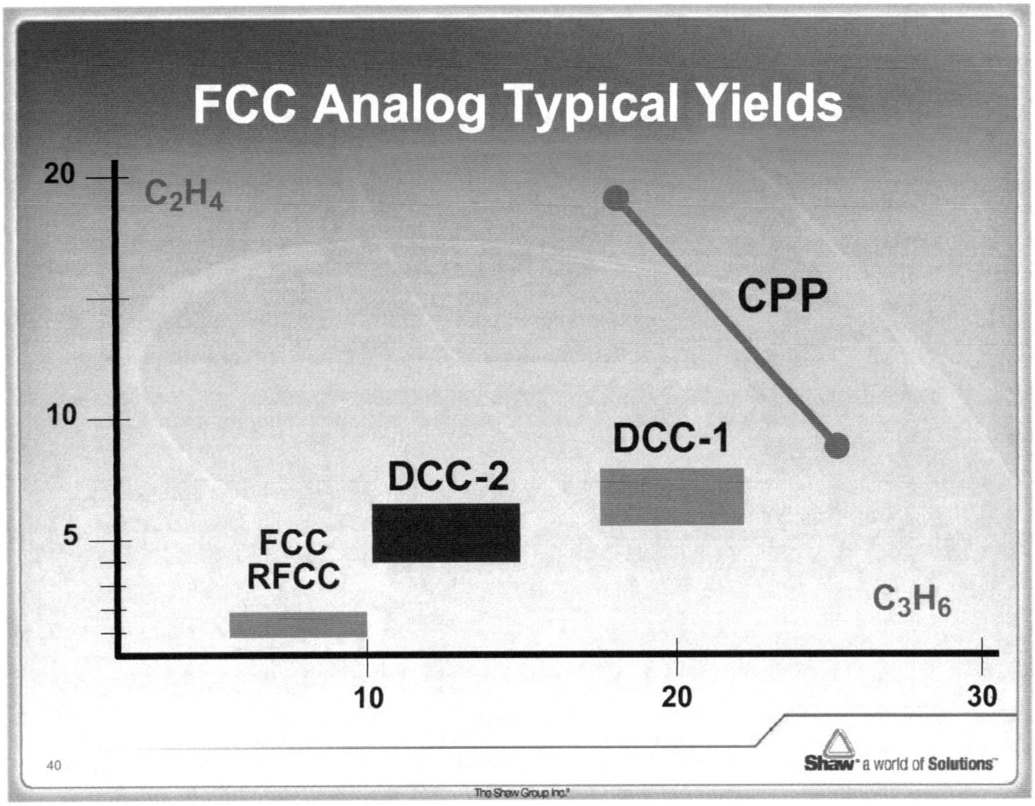
FCC Analog Typical Yields
20
10
5
C_2H_4
CPP
DCC-1
DCC-2
FCC
RFCC
C_3H_6
10
20
30
40
Shaw a world of Solutions
The Shaw Group Inc.

Olefin – Refinery Integration Conclusion

- Combined leading technologies – SC and FCC types
- Lower cost feedstocks – ethane, VGO/ATB
- Max ethylene – Catalytic Pyrolysis Process (CPP)
- Max propylene – Deep Catalytic Cracking (DCC)
- Increasing olefins – High Olefin FCC (HOFCC)
- Increasing olefins – recycle C_2, C_3, to SC; C_4, C_5 to HOFCC
- Lower investment cost, feedstock cost, energy cost

41

Shaw® a world of Solutions™

The Shaw Group Inc.®

Thank-You!

Questions?

42

Shaw® a world of Solutions™

The Shaw Group Inc.®

- **Ethylene Consultants – Ethylene – Linkage Solutions**

AIChE STS Galveston Conference
Advances in Petrochemical Technology
Olefin Co-product Links

Vasant Bhirud
1 October 2009

Vasant Bhirud - Background

- 2004 to date – Ethylene Consultants
  - BFLP/Port Arthur – Furnace Process & Engineering support
  - Consultations on acquisition of a cracker complex
  - Furnace run length validation
  - Cracker feedstock optimization study
- 2001 to 2004 UOP
  - Refinery – Ethylene – Polymer Complex : Feasibility study
  - Refinery – Ethylene Complex : Profit improvement study
  - Refinery – Ethylene – Petrochemical Complex : LP studies
- 1989 – 2001 Shaw Group – Stone & Webster

Vasant Bhirud October 1,2009 3

Olefin Co-product Links

- Proven technologies for improved olefins production
  - Aromatics extraction
  - Dehydrogenation
  - Isomerization
- Drivers
  - Hydrocarbon conservation to increase olefins yield
  - Economics

Ref: Zdonik et al, HP, Jan '76, page 149

Vasant Bhirud October 1,2009 4

Aromatics Extraction – Introduction

- Since aromatic species in naphtha have no olefin potential we have analyzed the economics of aromatics extraction using following cases.
  - One furnace performance and economics
  - Logistics of linkage for one furnace
  - Other advantages
  - Logistics of linkage on a cracker basis
  - Cracker performance and economics

Vasant Bhirud October 1,2009 5

Aromatics in a few Worldwide Naphthas

- Naphtha's are getting richer in aromatics
- List of full range naphthas

Naphtha	Country	Temperature range ^0C	Paraffins vol%	Naphthenes vol%	Aromatics vol%
Full Range Naphtha					
Ardjuna	Indonesia	65 - 193	34.30	46.00	19.70
Bombay Hi	India	C5 - 150	53.70	25.00	21.30
Dai Hung	Viet Nam	C5 - 175	37.64	30.35	32.01
Isthmus	Mexico	C5 - 175	56.00	21.90	22.10
Margham	Dubai	C5 - 165	56.16	20.73	23.11
Maya	Mexico	C5 - 175	40.70	32.40	26.90
Rangeland	Canada	C5 - 190	42.45	36.86	20.69
Statjford	North Sea	65 - 180	44.61	35.72	19.67
Tapis	Malaysia	C5 - 165	45.19	29.07	25.74

Vasant Bhirud October 1,2009 6

Naphtha Splitter

- Normally naphtha is split into light and heavy fractions in a refinery. All light naphtha (C_5 & lighter) and some heavy naphtha (C_6+) streams are sent to cracker

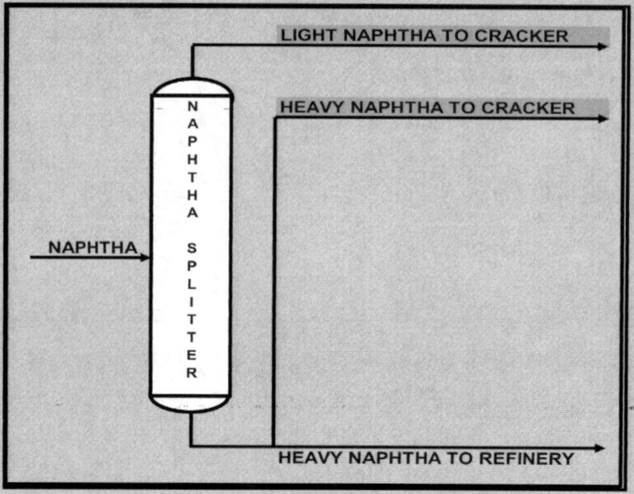

Vasant Bhirud October 1,2009 7

Worldwide Heavy Naphtha's

- Heavy naphtha's are specially rich in aromatics
- List of some worldwide heavy naphthas:

Naphtha	Country	Temperature range ^0C	Paraffins vol%	Naphthenes vol%	Aromatics vol%
Heavy Naphtha					
Attaka	Indonesia	100 - 150	41.00	34.30	24.70
Burgan	Divided Zone	93 - 223	44.00	36.00	20.00
Dai Hung	Viet Nam	95 - 175	33.60	30.80	35.60
Dukhan	Qatar	130 - 177	64.30	14.60	21.10
Qatar Marine	Qatar	129 - 177	56.70	20.40	22.90
Sharjah Conden	UAE	93 - 160	55.90	21.50	22.60

- Wt% content in naphtha is higher than vol %.
- Vol% content can be misleading

Ref: Bhirud, HP, April 2007

Vasant Bhirud October 1,2009 8

Furnace Performance of Heavy Naphthas

- Furnace performance comparison - cracking heavy naphtha "as is" vs "de-aromatized"
- Case study – Design naphtha vs operating naphtha (rich in aromatics) vs operating – dearomatized naphtha

	Heavy Naphtha Design	Heavy Naphtha Operating	Dearomatized operating naphtha
	vol%	vol%	vol%
N-P	30.60	27.20	34.00
I-P	36.74	32.66	40.82
Naphthenes	22.66	20.14	25.18
Aromatics	10.00	20.00	-
	100.00	100.00	100.00
SG	0.7281	0.7440	0.7123

- Ref: Bhirud, PTQ, Catalysis, 2008

Furnace Performance of Heavy Naphthas

Furnace Performance at Fixed Firing		Heavy Naphtha Design	Heavy Naphtha Operating	Dearomatized operating Naphtha
Propylene/Ethylene	Ratio	0.55	0.55	0.55
Feed rate	Ratio	1.000	1.08	0.92
Ethylene rate	Ratio	1.000	0.96	1.04
Firing per lb of ethylene	Btu/lb C2-	1.000	1.05	0.96
Coke laydown	Ratio	1.000	1.12	0.84
Run length fraction	Ratio	1.000	0.89	1.20
Run length fraction	Ratio		1.00	1.34
Furnace Ethylene capacity	lbs/hr	25,000	23,937	25,998
Ethylene production gain (loss)	lbs/hr		(1,063)	998
Ethylene revenue gain (loss), $0.3/lb	$/hr		(319)	300
Ethylene revenue gain (loss), $0.3/lb	$ MM/yr		(2.6)	2.4
Ethylene+Propylene revenue gain (loss)	$ MM/yr		(4.0)	3.7

- Ref: Bhirud, PTQ, Catalysis, 2008

Older Naphtha Cracker Configuration

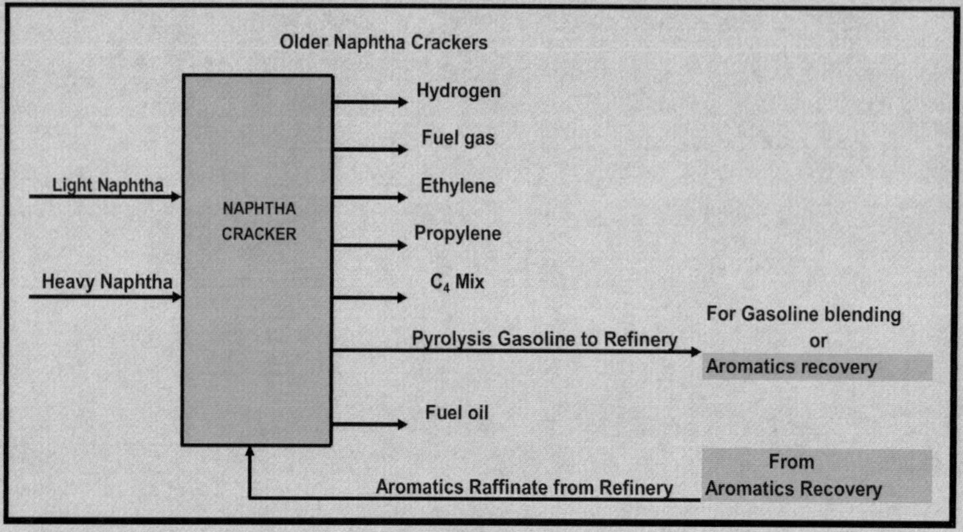

Vasant Bhirud October 1,2009 11

Newer Crackers with Aromatics Extraction

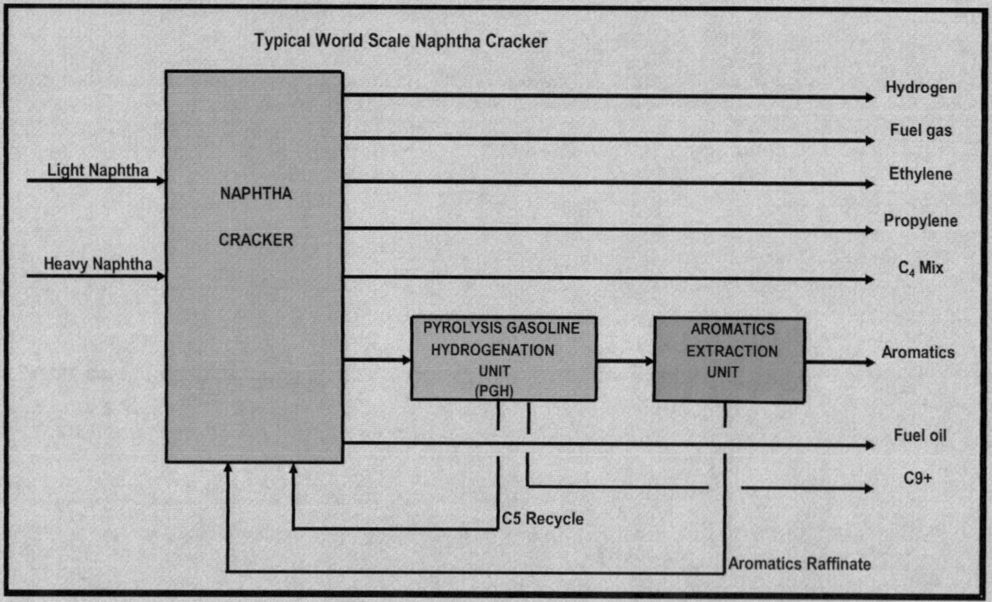

Vasant Bhirud October 1,2009 12

Send Heavy Naphtha for Aromatics Extraction

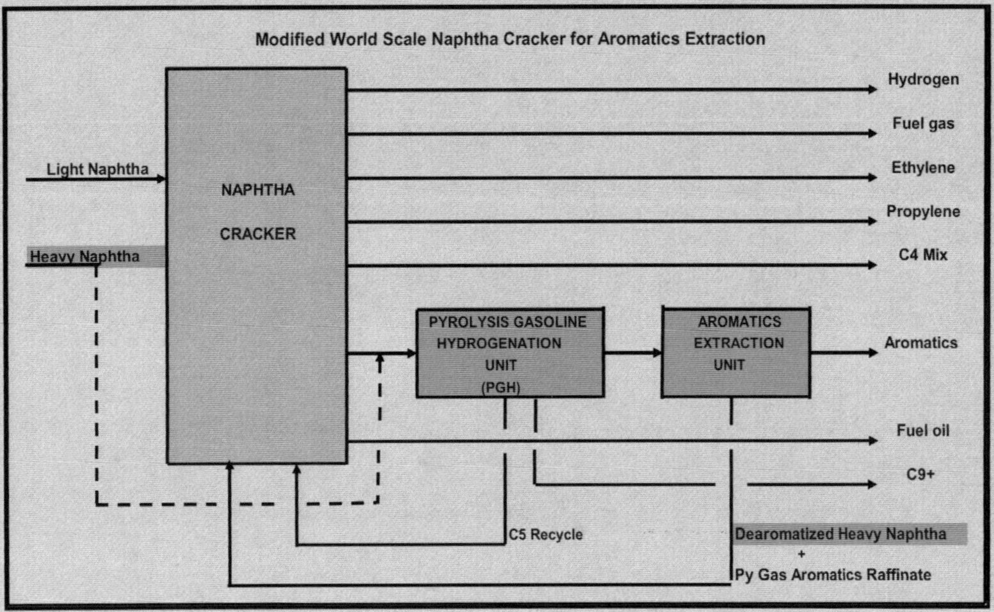

Vasant Bhirud October 1,2009 13

Economics – One Furnace Feed De-aromatized

- Economy of scale with one furnace feed

		Cracker	Cracker + 1 Heavy Nap Furnace
Py gas feed	kBPD	15	22.5
TIC : PGH + Arom Extr	$ MM	29	37
Delta TIC : PGH + Arom Extr	$ MM		8
Delta Revenues	$ MM/yr		7.7
Payout period	Years		1

Vasant Bhirud October 1,2009 14

Other Furnace Parameters

Furnace Performance at Fixed Firing

		Heavy Naphtha Design	Heavy Naphtha Operating	Dearomatized operating Naphtha
Propylene/Ethylene	Ratio	0.55	0.55	0.55
Feed rate	Ratio	1.00	1.12	0.84
Gasoline yield	Ratio	1.00	1.24	0.77
Fuel Oil yield	Ratio	1.00	1.52	0.41
Firing duty / lb of ethylene	Ratio	1.00	1.05	0.96

Furnace Performance - Summary

- Firing per lb of ethylene down
  - Fuel savings
- Coke lay down decreases
  - Furnace run length increases
  - Furnace availability increases
  - Yearly ethylene production increases due to availability
- Feed rate per lb of ethylene decreases
  - Dilution steam needs decrease
  - Debottlenecking of dilution steam circuit
- Sending heavy naphtha for aromatics extraction is economical on the basis of one furnace feed

Other Advantages of Naphtha De-aromatization

- Higher feed efficiency due to lower gasoline and fuel oil yield
- Savings in furnace fuel
- Increase in furnace run length
- Debottleneck dilution steam circuit
- Debottleneck quench oil tower
- Debottleneck quench oil circuit
- Debottleneck quench water tower
- Debottleneck quench water circuit
- Debottleneck distillate stripper

Vasant Bhirud October 1,2009 17

800 kTA Cracker Basis – Feed Cases

Naphtha Characterizations

	Base wt%	Naphtha A wt%	Dearom A wt%	Naphtha B wt%	Dearom B wt%
N-Paraffins	31.9	25.0	32.9	20.0	26.7
I-Paraffins	34.4	30.0	39.5	21.0	28.0
Naphthenes	24.3	21.0	27.6	34.0	45.3
Aromatics	9.4	24.0	-	25.0	-
	100.0	100.0	100.0	100.0	100.0
Total Paraffins	66.3	55.0	72.4	41.0	54.7
SG	0.6987	0.7230	0.6862	0.7417	0.7069

- Base naphtha is paraffin rich (> 65%)– suitable for cracker
- Naphtha A and B are paraffin poor – not suitable for cracker
- Upgrade naphtha A and B by de-aromatization

Vasant Bhirud October 1,2009 18

Cracker with Naphtha Hydrotreater (NHT)

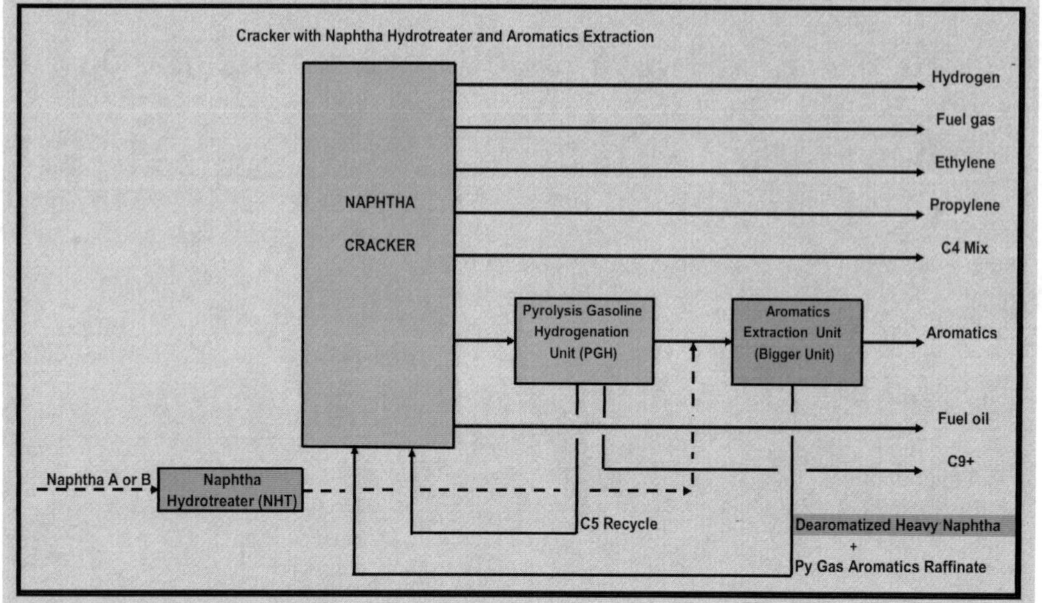

Vasant Bhirud October 1,2009 19

Plant Economics- De-aromatized A

		Base	Dearom-A
Naphtha			
Propylene/Ethylene ratio		0.5	0.5
Feeds	$/T	kTA	kTA
Base Naphtha	213	2,455.6	-
Dearomatized A or B	177	-	3,022.2
Total		2,455.6	3,022.2
Products			
Hydrogen	450	15.6	17.0
FG	180	389.7	405.7
Ethylene	372	800.0	800.0
Propylene	347	400.0	400.0
Butadiene	364	127.1	136.3
Raffinate -1	230	126.9	132.6
Benzene	253	174.9	221.3
Toluene	298	105.0	202.4
Xylenes+EB	299	88.2	295.7
C9+ Aromatics	290	103.3	329.7
FO	100	125.0	81.5
Total		2,455.6	3,022.2
Gross Margin	$miilion/yr	209.6	370.7
Delta Gross Margin	$miilion/yr	Base	161.1
Naphtha hydrotreater capacity	BPD		78,879
Aromatics extraction capacity	BPD	10,510	79,496
Naphtha hydrotreater cost	$ million	-	42.0
Aromatics extraction cost	$ million	10.5	43.4
Total cost	$ million	10.52	85.36
Delta cost	$ million	Base	74.83
Payout period	Years		0.5

Vasant Bhirud October 1,2009 20

Plant Economics – De-aromatized B

Naphtha			Base	Dearom -B
Propylene/Ethylene ratio			0.5	0.5
Feeds	$/T		kTA	kTA
Base Naphtha		213	2,455.6	-
Dearomatized A or B		177	-	3,240.7
Total			2,455.6	3,240.7
Products				
Hydrogen		450	15.6	17.4
FG		180	389.7	411.1
Ethylene		372	800.0	800.0
Propylene		347	400.0	400.0
Butadiene		364	127.1	151.5
Raffinate -1		230	126.9	131.5
Benzene		253	174.9	281.2
Toluene		298	105.0	238.7
Xylenes+EB		299	88.2	331.1
C9+ Aromatics		290	103.3	376.1
FO		100	125.0	102.0
Total			2,455.6	3,240.7
Gross Margin	$miilion/yr		209.6	390.5
Delta Gross Margin	$miilion/yr		Base	180.9
Naphtha hydrotreater capacity	BPD			84,583
Aromatics extraction capacity	BPD		10,510	88,087
Naphtha hydrotreater cost	$ million		-	43.8
Aromatics extraction cost	$ million		10.5	46.6
Total cost	$ million		10.52	90.38
Delta cost	$ million		Base	79.86
Payout period	Years			0.4

Vasant Bhirud October 1,2009 21

Plant Economics of De-aromatization

- De-aromatization of poor naphtha A or B is economical
- Payout period is about 0.5 years
- Poor naphtha with paraffin content as low as 41 wt% and aromatics content of 25 wt% can be processed in the existing cracker without loss of capacity
- If forced to process naphtha A or B, plant will be furnace short
- To process naphtha A or B to achieve design nameplate capacity will need to add a few furnaces and modify front end hot section
- Best alternative is to de-aromatize poor naphthas without installing additional furnaces

Vasant Bhirud October 1,2009 22

Aromatics Extraction Licensors

- Technology licensors are
  - UOP
  - Kruppe - Uhde
  - GTC
  - Axens

Propane Dehydrogenation

Propane Dehydrogenation-Introduction

- To maximize propylene production, many crackers are being designed or operating with propane as a significant portion of the feed. We have evaluated if linkage with propane dehydrogenation can be justified using following cases.
  - Ethane/Propane cracker with PDH
  - Propane/Naphtha cracker with PDH
  - PDH linkage with existing adjacent crackers

Vasant Bhirud October 1,2009 25

Ethane/Propane Cracker

- Some crackers are designed to maximize propylene production with Ethane/Propane mix feed as follows:

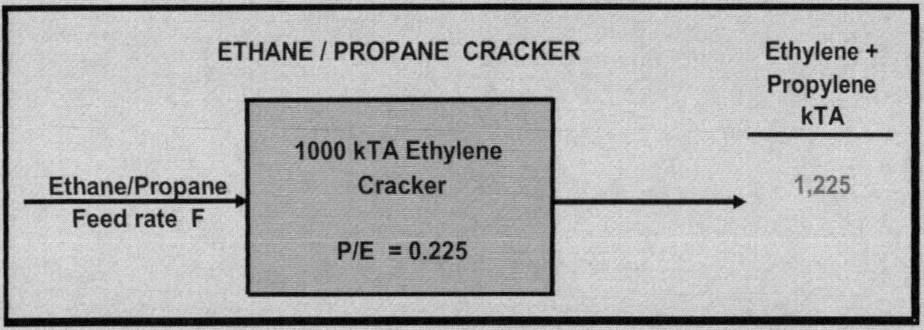

Vasant Bhirud October 1,2009 26

Ethane/Propane Cracker Strategy

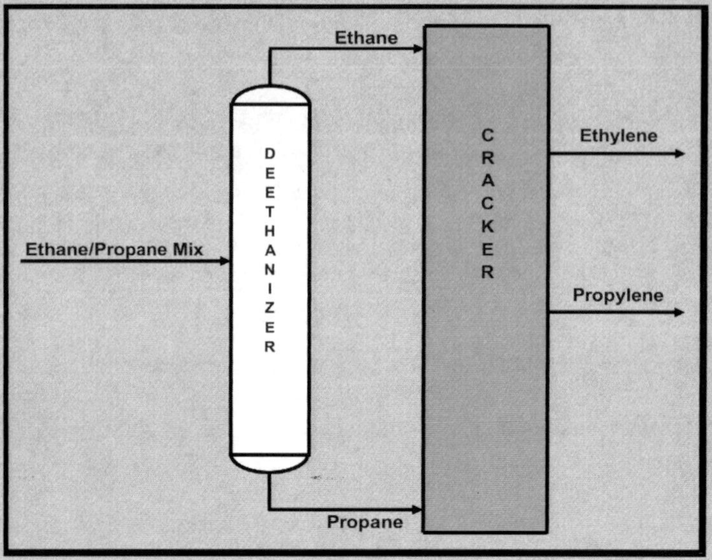

Vasant Bhirud October 1,2009 27

Propane Dehydrogenation (PDH) Licensors

- Propane dehydrogenation licensors
  - C_3 Oleflex - UOP
  - CATOFIN – CBI-LUMMUS
  - PROPYLUR (KBR)
  - KRUPPE
- This analysis based on UOP technology
- Other technologies could offer similar advantages

Vasant Bhirud October 1,2009 28

Ethane Cracker + Propane Dehydrogenation

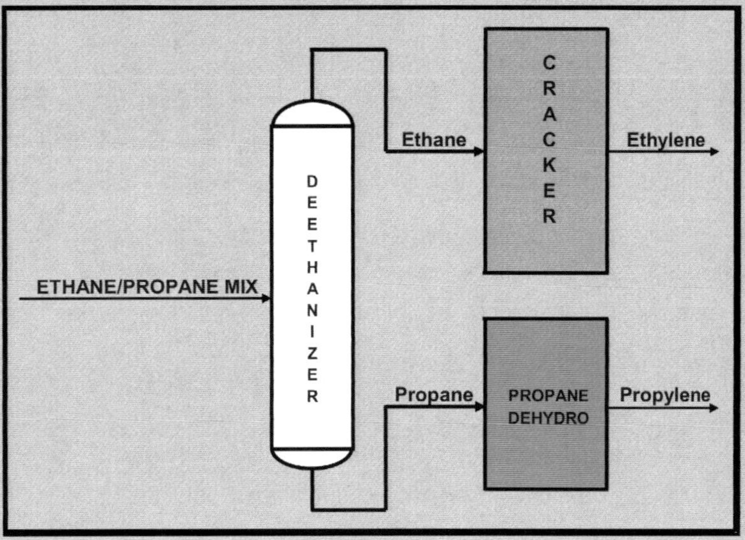

Vasant Bhirud October 1,2009 29

Ethane Cracker + Propane Dehydrogenation

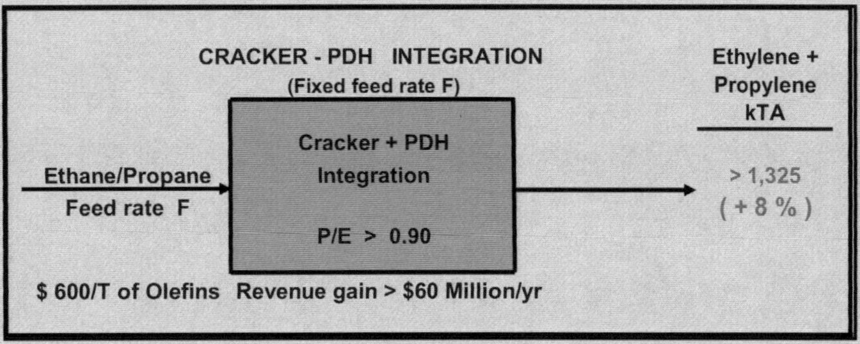

Vasant Bhirud October 1,2009 30

Economics of Ethane Cracker + PDH

		E/P Cracker	Cracker + PDH
TIC	$ MM	460	550
Delta TIC	$ MM		90
Ethylene + Propylene	kTA	1,225	1,325
Revenues @ $600/T	$ MM/yr	735	795
Delta Revenues	$ MM/yr	Base	60
Payout period	Years		1.5

Vasant Bhirud October 1,2009 31

Ethane Cracker + PDH - Summary

- Payout period for Ethane Cracker with PDH for E/P mix is about 1.5 years which can be justified when propylene production needs to be maximized
- No credit is taken for by-products
- No credit is taken for utility reduction

Vasant Bhirud October 1,2009 32

Propane/Naphtha Cracker + PDH

- **Some Propane/Naphtha crackers are designed to maximize propylene production**
- **In this case naphtha cracker can be linked to propane dehydrogenation unit to further increase propylene production significantly**

Vasant Bhirud October 1,2009 33

Propane/Naphtha Cracker + PDH

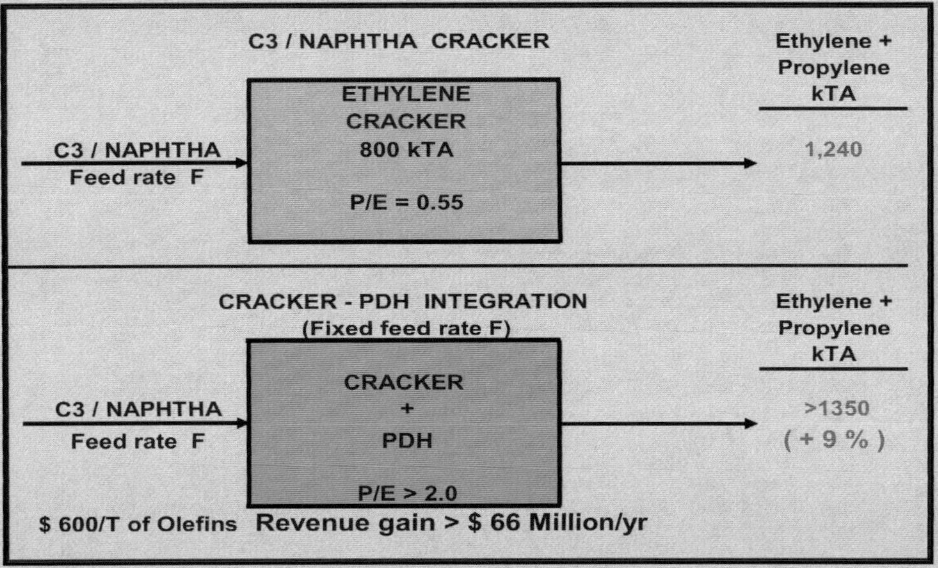

Vasant Bhirud October 1,2009 34

Cracker + Existing PDH

- **Some stand alone PDH units have been built around the world. By linking PDH units to an <u>adjacent</u> cracker, significant increase in propylene production can be achieved.**
- **Cost of linking will depend on the proximity of PDH with cracker**
- **Since most of the PDH and cracker units are built in petrochemical zones, their linking cost could be low**

Vasant Bhirud October 1,2009 35

Ethane Cracker + Existing PDH

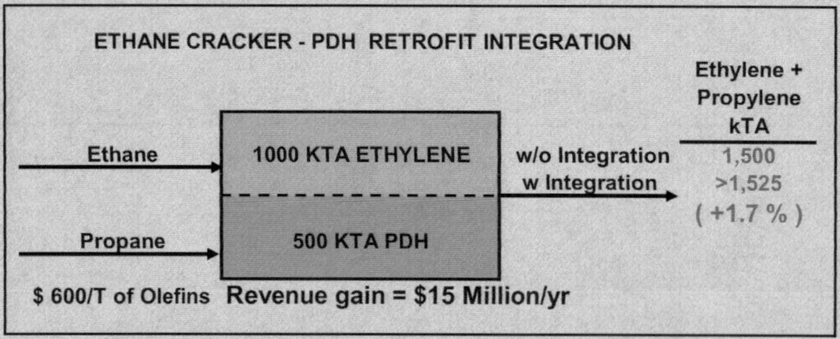

Vasant Bhirud October 1,2009 36

Naphtha Cracker + Existing PDH

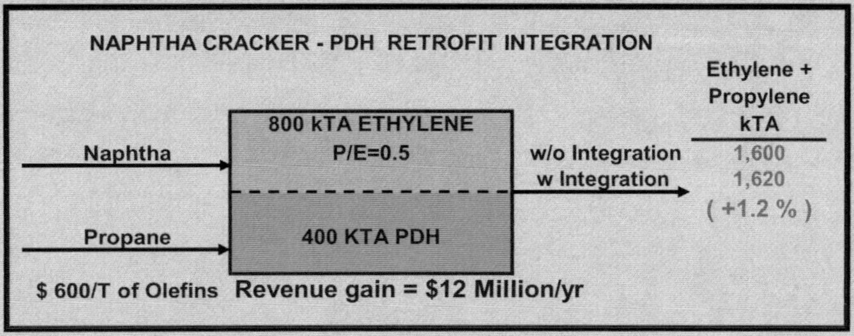

Isomerization

Isomerization - Introduction

- Since many crackers are fed with increasing quantities of I-Butane in butane mix feeds we evaluated isomerization of I-Butane to N-Butane using following cases:
  - Overall impact on plant capacity and economics
  - Furnace performance
  - Economics of isomerization unit with cracker

Isomerization of I-C4 to N-C4

- Basis - An ethylene cracker based on pure N-Butane feed
- N-Butane is probably supplied from refinery
- Cracker is likely to receive mixture of N and I butanes
- I-Butane cracker feed is very poor compared to N-Butane
- Is option to convert I-Butane to N-Butane justified?

Feed Cases for I-C4 Isomerization

- **Feed cases**

Case		1-A	1-B
Feed composition			
N-Butane	wt%	100	75
I-Butane	wt%	0	25
		100	100

Vasant Bhirud October 1,2009 41

Ethylene Cracker – N-Butane vs Mixed Butanes

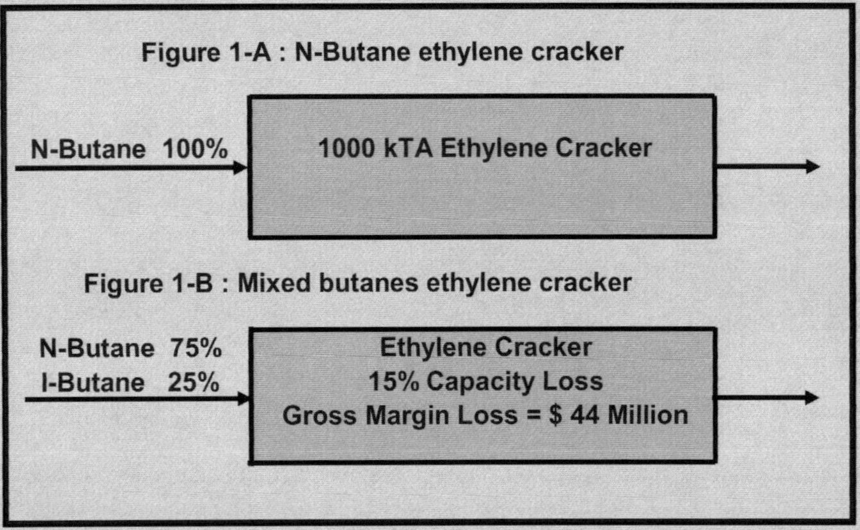

Mixed butane cracker – loss in capacity

Vasant Bhirud October 1,2009 42

Furnace Economics

- **Furnace performance**

Furnace Performance at Fixed Firing		N-C4 100%	N-C4/I-C4 75/25 %
N-Butane conversion	wt%	90	90
Feed rate	Ratio	1.000	1.03
Ethylene production	Ratio	1.000	0.86
Firing per lb of ethylene	Btu/lb C2-	1.000	1.16
Ethylene + Propylene	Ratio	1.000	0.936
Furnace Ethylene capacity	lbs/hr	25,000	21,482
Ethylene+ Propylene production	lbs/hr	35,484	33,207
Ethylene + propylene production (loss)	lbs/hr		(2,277)
Ethylene+Propylene revenue gain (loss) $0.30/lb	$ MM/yr		(5.5)

Vasant Bhirud October 1,2009 43

Isomerization with Ethylene Cracker

Figure 1-C - Mixed butanes - Feed isomerization integrated with cracker

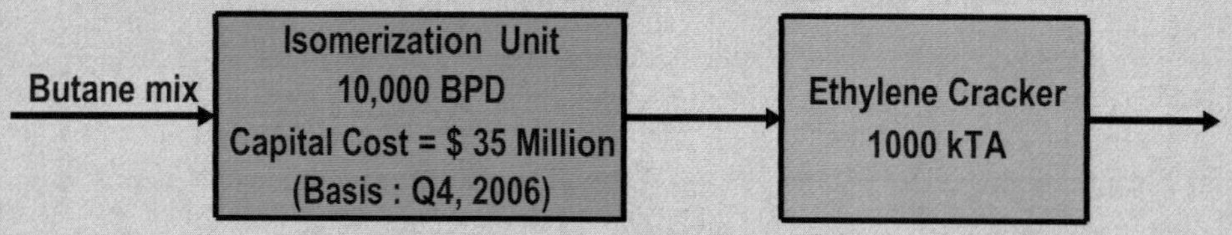

Vasant Bhirud October 1,2009 44

Plant Economics

			N-C4	N/I C4- 75/25
	C2- + C3-	kTA	1,390.1	1,317.0
		$/T	kTA	kTA
Feed	N-Butane/Mix		1,917.5	1,890.7
Products	Hydrogen	500.0	26.9	22.4
	FG	180.0	380.2	413.3
	Ethylene	600.0	1,000.0	869.4
	Propylene	460.0	390.1	447.6
	Py Gas NA	225.0	43.6	55.3
	Benzene	344.0	41.2	43.3
	Toluene	227.0	8.4	12.0
	Xylenes+EB	230.0	8.2	9.4
	C9+ Aromatics	232.0	3.7	3.9
	FO	140.0	15.3	14.2
			1,917.6	1,890.7
	Product Revenues	$ MM/yr	892.08	848.20
	Delta Prod Revenues	$ MM/yr	Base	(43.9)
	Isomerization Capacity	kBPD		10
	TIC	$MM		35
	Payout period	Years		0.8

Vasant Bhirud October 1,2009 45

Summary : I-C4 Isomerization

- I-Butane isomerization to N-Butane can be economical
- Isomerization is a proven technology
- Normally N-Butane is converted to I-Butane in refineries. This process can be modified to convert I-Butane to N-Butane
- Ethylene plant operator should be concerned about I-Butane in the feed, and its impact on furnace operations and cracker as a whole

Vasant Bhirud October 1,2009 46

Conclusions

- Proper olefin links can be economical
- They offer "flexibility" to ethylene operators
- Smart "flexibility" will be key to survival

- Additional economic improvements are possible depending on locations
- Ethylene consultants can work with clients and technology licensors to properly link these technologies with cracker

Vasant Bhirud October 1,2009 47

Acknowledgements

- Shaw Group – All photos
- Pycos Model – Ethylene yields
- UOP – PDH - General yield vector

- Reference : Bhirud, HP, April 2007
- Reference : Bhirud, PTQ - Catalyst, 2008

Vasant Bhirud October 1,2009 48

Cracker Links

Vasant Bhirud October 1,2009 49

END

- **THANK YOU**

- **QUESTIONS ?**

Vasant Bhirud October 1,2009 50

Challenges in Developing an Advanced Post-Combustion Capture Technology

2009 AIChE Regional Process Technology Conference – Galveston, Texas

Mr. Kumar Sugavanam – BASF Corporation
Dr. Georg Sieder – BASF SE
Dr. Andreas Northemann – BASF SE
Mr. Hugo Garcia – BASF SE

BASF
The Chemical Company

Carbon Capture and Storage (CCS)

Options for fossil fired power plants

BASF
The Chemical Company

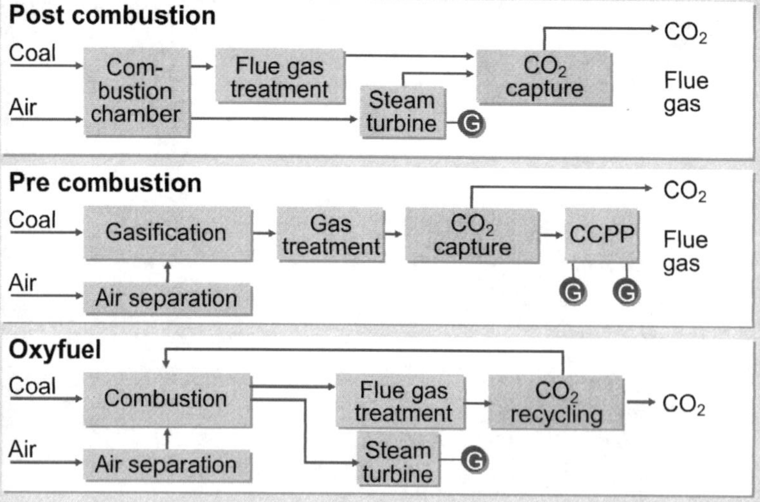

Carbon Capture and Storage (CCS)
Principle of Post-Combustion Capture

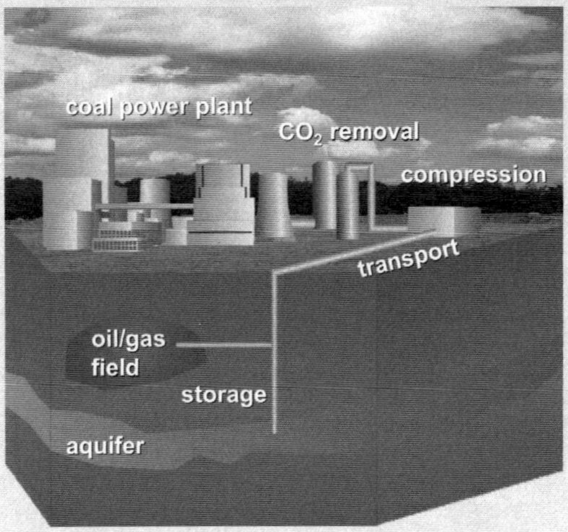

- Post combustion capture (especially chemical absorption):
  - Capture ready plants
  - Proven technology
  - Can be used for existing power plant technology

- BASF has long-term experience in gas treatment for syngas and natural gas (including CCS)
- More than 250 reference plants for the aMDEA® technology

Post-Combustion Capture
Process Scheme

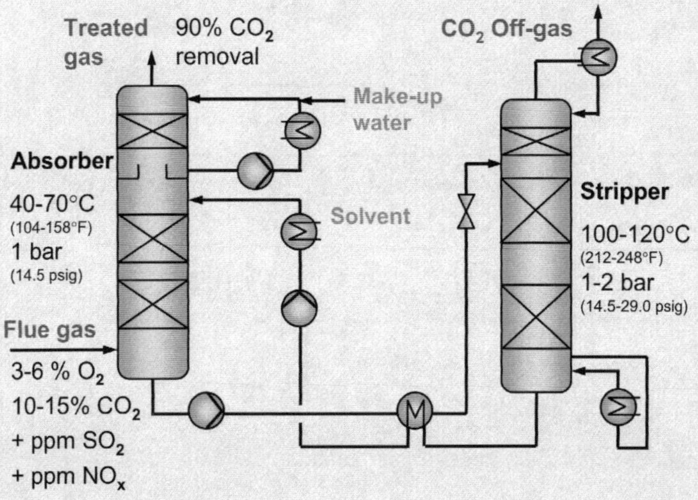

Additionally:

- Blower
- *Pre-wash/ Direct Cooler*
- Reclaimer
- Compressors

Major contributing factors to CCS costs of a PCC technology

BASF
The Chemical Company

Capture Plant

Equipment size

Absorber/Stripper (height and diameter)

Pumps

Process configuration

Material of construction

Power plant integration

Solvent

Energy requirement

Capacity

Kinetics

Stability

Objectives of BASF's development program

BASF
The Chemical Company

Solvent	MEA (30%)	BASF Process
Power plant efficiency loss	12-14 % points	< 10 % points
CAPEX	100 %	80 %
Product losses	1.6 – 3.0 kg/t CO_2	< 0.4 kg/t CO_2

Timeline – PCC Technology R&D

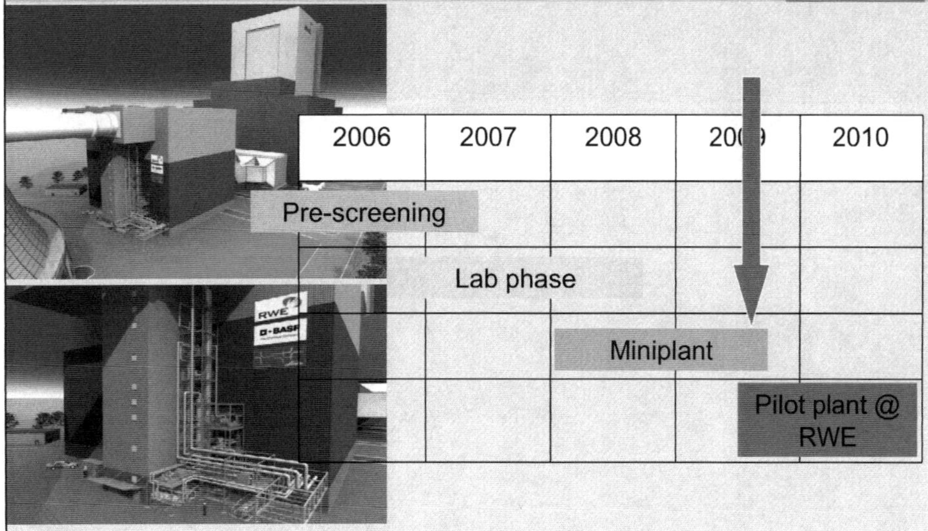

The long road to success
Solvent Screening Procedure for PCC

Product portfolio, databases ~ 400 Substances

Pre-selection: Molecular weight, Vapour pressure, Alkalinity, MSDS

~ 180

Screening: Phase equilibria, Reaction kinetics, Stability — **Discontinuous Synthetic gas**

10 - 20

Verification: Mini plant — **Continuous Synthetic gas**

2

Litmus test: Pilot plant — **Continuous Slip stream**

Selected Results of the Mini plant runs

Mini plant at BASF

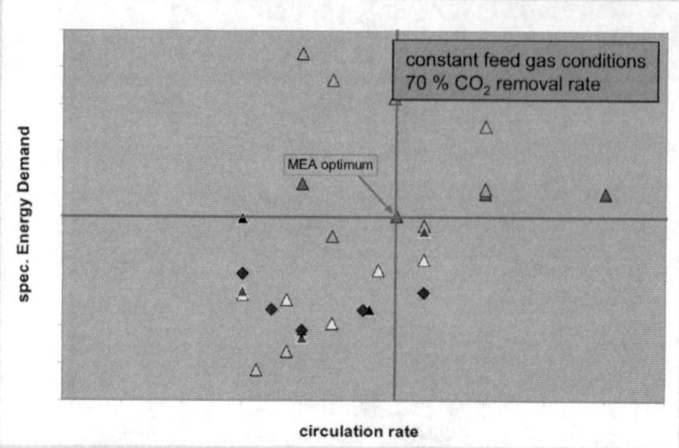

- Verification of the screening results
- Several options for an advanced solvent are identified

Pilot Phase and Full Scale Studies RWE – Linde – BASF

Work packages
Pilot plant
Optimization
Full scale plant (study)

Goals
CO_2 avoidance cost < CO_2 price
Efficiency loss < 10 %-point

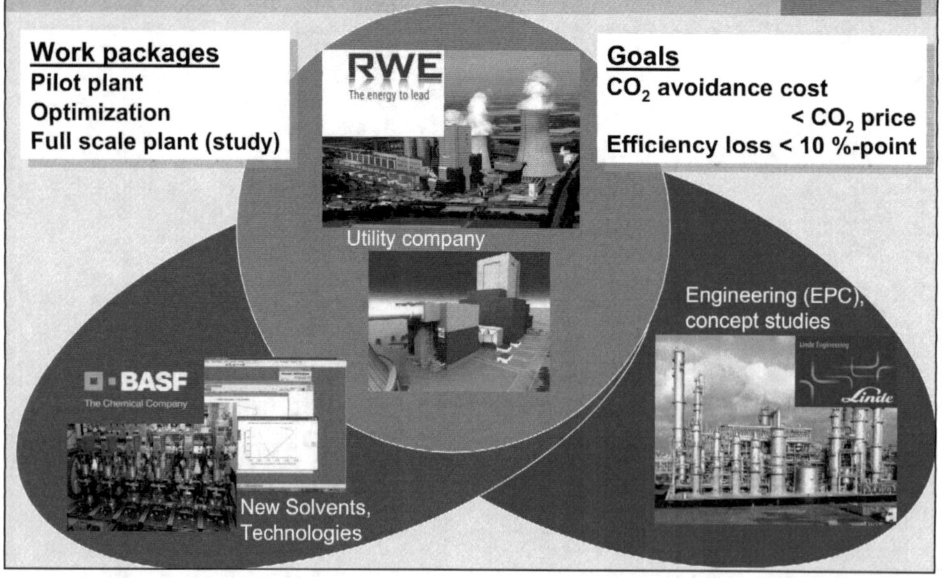

The Niederaussem Pilot Plant
Technical facts

BASF
The Chemical Company

- Type: Pilot plant at a lignite fired power plant
- Flow rate: 1550 Nm^3/hr flue gas (1.4 MMSCFD)
- CO_2-capacity: 7.2 tpd (661 lb/hr)
- Design: 6 skids, 2 container, max. height 41 m
- Time-line: start-up in June 2009
  test runs until end of 2010 with MEA (as benchmark) and 2 new solvents
- Test program: solvents under real conditions
  regeneration energy
  degradation rate
  solvent losses
  material tests in the pilot plant
  240 different measurements

The Niederaussem Pilot Plant
Process Flow Sheet

BASF
The Chemical Company

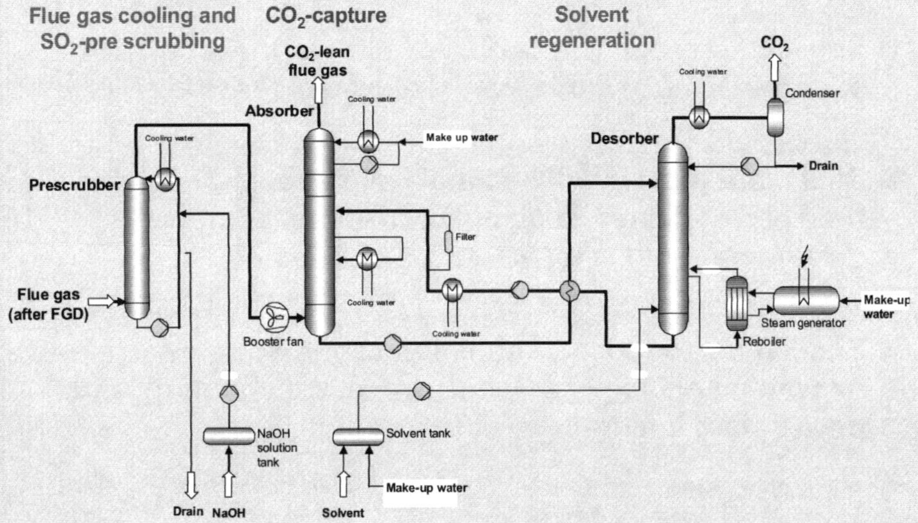

Question of Scale Up
Study jointly with RWE and Linde

BASF
The Chemical Company

Pilot plant	Demo plant	Commercial plant
approx. 6-10 Mill €	approx. 60-190 Mill €	approx. 500 Mill €
0.3 t CO_2 / hr	35 - 130 t CO_2 / hr	750 t CO_2 / hr
1550 Nm^3/h flue gas	140 – 680*10^3 Nm^3/h flue gas	3*10^6 Nm^3/h flue gas
0.45 MW_{el}	50 - 250 MW_{el}	1100 MW_{el}

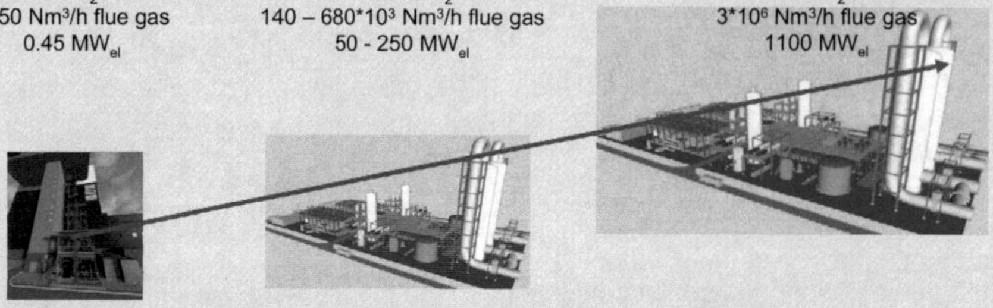

2008 2009 2010 2011 2012 2013 2014 2015 2016 2017 2018

Goals:

- **Validation of technology until 2015 in demo plant**
- **Commercial availability before 2020**

BASF CONFIDENTIAL 13

Summary

BASF
The Chemical Company

- **Solvent selection is a decisive factor in the long term success of CCS; more than 170 systems tested**
- **Pilot plant trials in 2009 using a slip stream of lignite fired power plant flue gas plan to verify laboratory data on newly developed solvent technology**
- **Economics for post combustion capture largely depend on choice of solvent, process technology and power plant integration**

BASF CONFIDENTIAL 14

Overview of Advances in EB/Styrene and Phenolics Technologies

Vincent Welch
Global Licensing Manager, Styrenics
Badger Licensing LLC

STS Galveston
October 1, 2009

Badger Licensing LLC - Overview

- **Joint venture between The Shaw Group and ExxonMobil Chemical, formed 2003**
- **Legacy business of the Badger Company and its successors**
- **Develops, licenses and commercializes process technologies**
- **Headquartered in Cambridge, MA**
- **Performs R&D in Weymouth, MA**

Badger
Licensing LLC

Principal Technologies

- **Phenolics Chain**
  - Cumene
  - Bisphenol A
    - Advanced co-promoted system
    - Promoter bound catalyst
- **Styrenics Chain**
  - Ethylbenzene
    - Liquid Phase, Mixed Phase, Vapor Phase
  - Styrene
    - Phenylacetylene Reduction
    - Catalyst Stabilization

Badger
Licensing LLC

Technology Portfolio

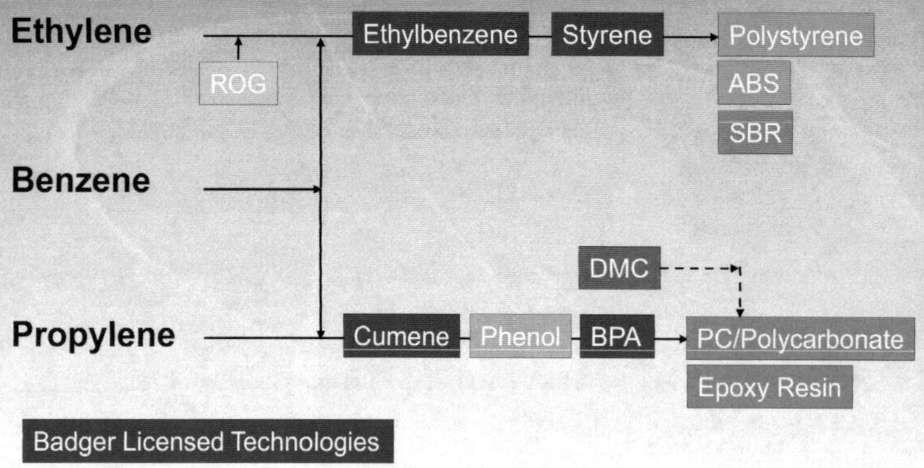

Badger
Licensing LLC

Licenses – Last 15 Years

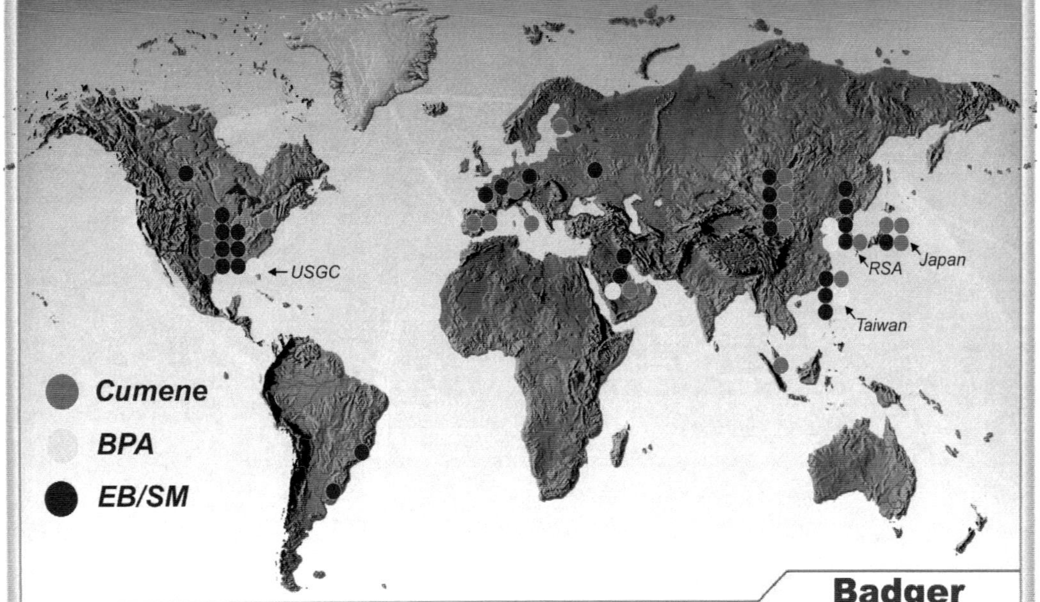

Badger
Licensing LLC

Weymouth Research Center

- **Shaw R&D Center - more than 30 years**
  - Located 20 miles south of Boston
- **Operating pilot units including**
  - EB
  - SM
  - Cumene
  - BPA
  - Benzene Reduction
- **Focus is process development and commercialization**
  - Maintain and improve proprietary technologies

Badger
Licensing LLC

BPA Technology

- **Technology platform developed by Shell (acquired by RPP/Hexion)**
  - Resin Based Technology
  - Derived from more than 50 years of continuous BPA production experience
  - Long history of BPA production in USA and Europe
- **Badger Licensing acquired technology rights in December 2004**
  - Licensed five times, all large plants
  - 50% share of licensed market
  - Dominant product in merchant BPA market

First Licensee – Chang Chun in Kaohsiung, Taiwan

Badger Licensing LLC

BPA Process

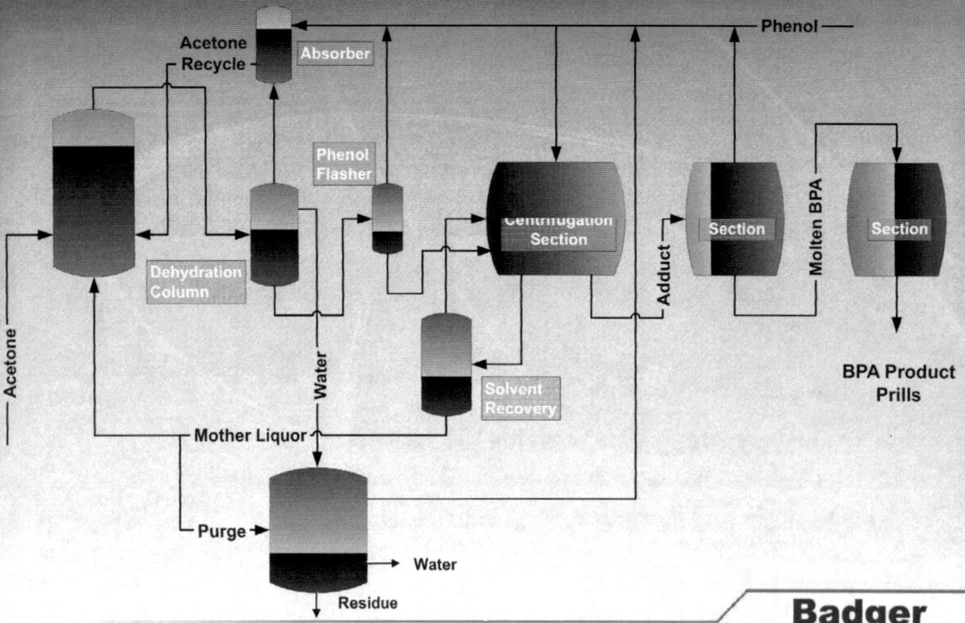

Badger Licensing LLC

BPA Technology Advantages

- **Largest capacity available for license in single train**
- **Very high purity product, excellent color**
  - Samples are available for prospective licensee
- **All major polycarbonate producers/technologies have used Badger's BPA product**
- **Process and mechanical design results in high reliability and on-stream factor**
  - Unplanned outages less than 5 days per year
  - +98% on-stream factor
- **Long cycle time between major turnarounds**
  - 10 years achieved

Badger Licensing LLC

BPA Commercial Experience

Company	Location	Capacity MTA	Startup Date
Shell (now Hexion)	Deer Park, TX, USA	143,000	1992
LG Petrochemical	Republic of Korea	120,000	2005
Chang Chun	Taiwan	135,000	2005
PMD (now Kayan)	KSA	240,000	2010
LG Chem	Korea	120,000	2008
Confidential	Asia	135,000	2009

Badger Licensing LLC

Cumene Technology

- **Badger's technology has displaced vast majority of cumene produced by older technologies**
  - Solid Phosphoric Acid (SPA)
  - Aluminum Chloride ($AlCl_3$)
- **Features ExxonMobil's proprietary zeolite catalysts specifically designed for the production of cumene**
  - Selectivity to cumene surpasses that of common zeolites
  - Equally effective for transalkylation of polyalkylated benzenes to cumene
  - Minimum byproduct formation provided high yield and ultra high product purity

Badger
Licensing LLC

Cumene Milestones

- **1992 - Technology introduced with a B/P ratio of 4:1 molar, Bz/PIPB ratio of 3:1 weight**
  - Significant capacity increase for SPA plants
- **1996 - Alkylation B/P ratio reduced to 3:1 molar**
  - Additional capacity increase for SPA and $AlCl_3$ plants
- **2009 - Technology commercially operating at B/P ratios as low as 1.8:1 and Bz/PIPB ratios of 1:1**
  - Maintained excellent product purity
- **Advancements in catalyst**
  - Second generation alkylation catalyst commercialized
  - Second and third generation catalyst commercialized for transalkylation

23 Licensed Plants

Badger
Licensing LLC

Cumene Advantages

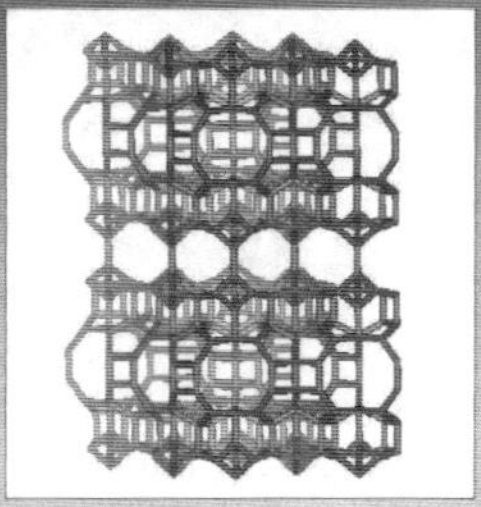

- **Badger's cumene process features:**
  - Low reactor operating temperatures
  - Low catalyst loading
  - Low B/P and Bz/PIPB ratios
- **ExxonMobil's catalyst does not produce oligomers and does not age through coking**
  - Catalyst cycle length not affected by B/P ratio
  - Catalyst stability at very low B/P ratios commercially demonstrated
- **Badger's cumene technology is now routinely being offered at extremely low B/P ratios, unmatched by any other cumene technology**

Badger
Licensing LLC

Cumene Process

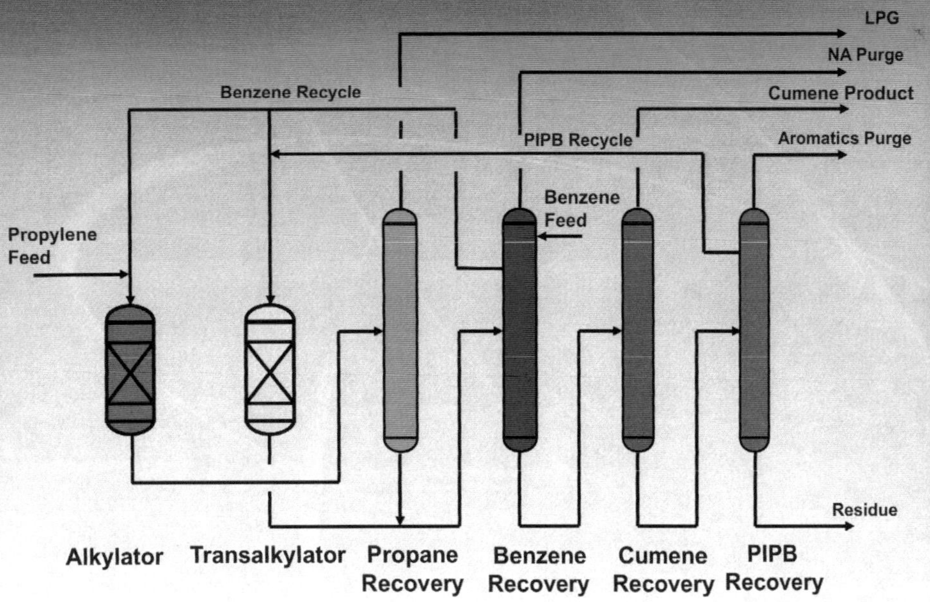

Badger
Licensing LLC

Cumene Commercial Experience

- **Licensed 23 plants**
  - 18 licensees operating with 6,600,000 MTA combined capacity and over 125 years of combined operation
  - 5 licensees in design represents additional 1,215,000 MTA
- **Commercially demonstrated long alkylation catalyst cycle lengths**
  - First cycle of over 8 years at B/P ratio of 3:1 using refinery grade propylene
  - First cycle of over 7 years (so far) at B/P ratio of 2.5:1
  - First cycle of nearly 4 years (so far) at B/P ratio of 1.8:1
- **No transalkylation catalyst has required regeneration**

Badger Licensing LLC

Cumene Process Flexibility

- **Feedstock flexibility**:
  - Polymer grade – 6 plants
  - Chemical grade – 5 plants
  - Refinery grade – 9 plants
  - Polymer or chemical grade – 2 plants
  - Polymer or refinery grade – 1 plant
- **Design experience:**
  - SPA revamps – 8 plants
  - $AlCl_3$ revamps – 4 plants
  - Grassroots – 11 plants

Badger Licensing LLC

Experience Summary

	Location	Project Type	Capacity (MTA)
Georgia Gulf	USA	SPA Revamp	680,000
Flint Hills	USA	SPA Revamp	640,000
Borealis	Finland	$AlCl_3$ Revamp	245,000
Citgo	USA	SPA Revamp	690,000
Caproleuna	Germany	Grassroots	245,000
Ertisa (CEPSA)	Spain	SPA Revamp	430,000
Sun Oil	USA	SPA Revamp	500,000
Shell Chemical	USA	$AlCl_3$ Revamp	725,000
Mitsui Chemical	Singapore	Grassroots	340,000
Mitsubishi	Japan	SPA Revamp	340,000
Novapex	Europe	$AlCl_3$ Revamp	280,000
Mitsui Chemical	Japan	SPA Revamp	280,000
LG Petrochem	Korea	Grassroots	200,000
Taiwan Prosperity	Taiwan	$AlCl_3$ Revamp	240,000
SGQPC	China	Grassroots	167,000
Ertisa (CEPSA)	Spain	Grassroots	300,000
Kingboard	China	Grassroots	167,000
NPRC	Japan	Grassroots	170,000
Saudi Kayan	Saudi Arabia	Grassroots	290,000
Undisclosed	Europe	SPA Revamp	300,000
Undisclosed	China	Grassroots	300,000
Bluestar	China	Grassroots	325,000
Undisclosed	Asia	Grassroots	400,000

Badger
Licensing LLC

EB/SM Technologies

- **Leading supplier of Ethylbenzene/Styrene Monomer (EB/SM) technology since 1964**
- **Partners are critical to technology development and maintenance**
  - EB: ExxonMobil
  - Styrene: TOTAL Petrochemicals
- **Committed to technology development**
  - State-of-the-art pilot plants
  - On-going engineering programs
- **Badger and partners provide support to licensees throughout plant life**
  - Assures plant efficiency is maintained

Badger
Licensing LLC

EBMax Technology

- **Liquid or partial liquid phase alkylation**
  - Polymer grade or dilute ethylene feedstock
  - ExxonMobil EBMax alkylation catalyst
- **Liquid phase transalkylation**
  - ExxonMobil EBMax transalkylation catalyst
- **Commercialized in 1995 at Chiba SM**
- **31 plants licensed, 16.2 MM MTA capacity**
- **Distinguishing process features**
  - Operates with low excess of benzene to reactors
  - Operates over a wide range of temperatures
  - Low alkylation and transalkylation catalyst loads
  - Long catalyst cycle lengths

Badger
Licensing LLC

Ethylbenzene Milestones

- **Before 1980, most EB produced by $AlCl_3$**
- **Vapor Phase process commercialized in 1980**
- **Vapor Phase process using RFCC Off-gas Feed**
  - Commercialized at Shell Stanlow in 1991
  - Second plant – ChevronPhillips in 1999
- **EBMax process commercialized in 1995**
  - Alkylation in liquid phase over EM MCM-22 catalyst
- **Second generation catalysts introduced in 2004**
  - Catalysts with higher activity and selectivity
  - Equipment sizes, raw material consumption, and energy consumption all reduced

65 Licensed Plants

Badger
Licensing LLC

EBMax Technology Advantages

- **Operates with low B:E feed ratios**
  - Reduces capital cost of reaction and benzene recovery systems
  - Two EBMax plants operating at B:E < 1.7:1 molar
- **Catalyst is stable at low bed inlet temperatures**
  - Widens operating temperature envelope
  - Reduces reactor cost (fewer catalyst beds)
- **Low alkylation reactor operating pressure**
  - Reduces reactor cost (lower design pressure)
  - Reduces ethylene supply pressure requirement
    - Supply pressure as low as 26 kg/cm^2(g) at reactor is adequate
- **Low alkylation catalyst load**
  - Allowance for activity loss due to coking not required

Badger
Licensing LLC

EBMax Alkylation Catalyst

- **Developed specifically by EM for benzene alkylation**
- **Adsorption of benzene preferred over ethylene**
  - Minimizes reaction of ethylene with itself (oligomerization)
  - Virtually eliminates undesirable side reactions
    - Alkylaromatic impurities in EB product (cumene)
    - Non-aromatics (cycloparaffins such as cyclohexane)
  - Operates with stability at low operating temperatures

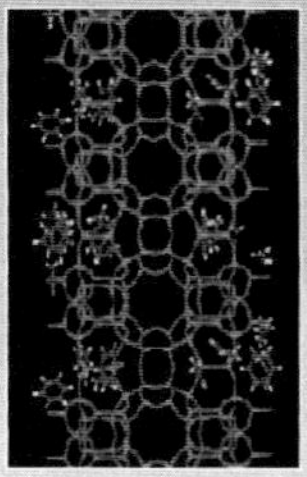

- **Catalyst does not age by "coking"**
- **Excess benzene not required to maintain activity**
  - Only to moderate reaction exotherm
- **Monoalkylation preferred over polyalkylation**
  - Minimizes PEB production

Badger
Licensing LLC

EBMax Process

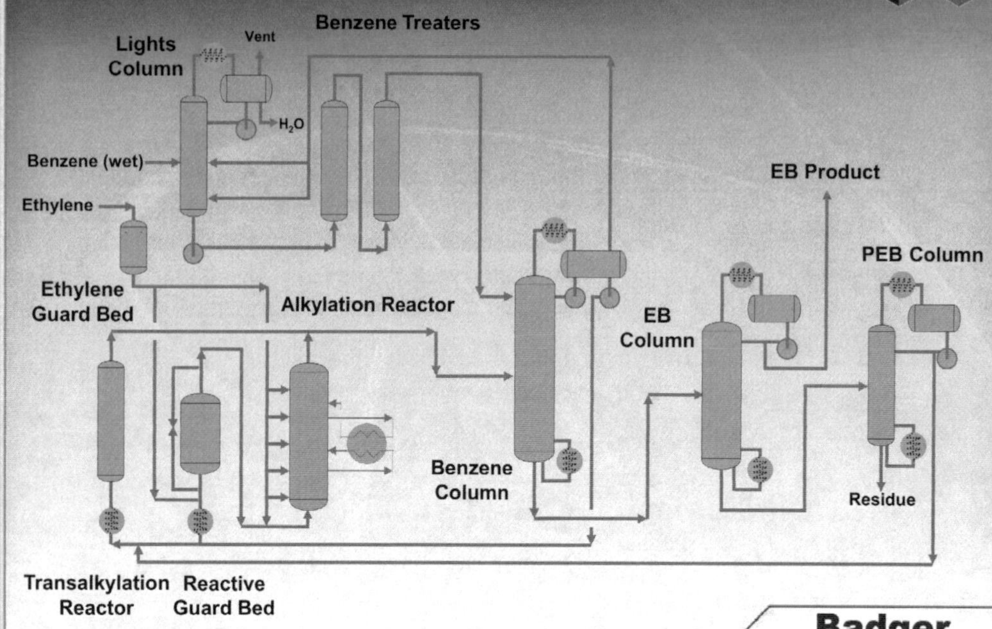

Badger
Licensing LLC

Reactive Guard Bed

- **Pretreatment does not remove 100% of N compounds**
  - Some compounds adsorbed more strongly than others
  - Those not removed adsorb on first alkylation catalyst bed
- **First alkylation bed contained in separate vessel**
  - Reactive Guard Bed (RGB)
  - RGB provides second line of protection
  - Protects catalyst in alkylation reactor
    - more effective than mole sieve in removing final traces
- **Remove RGB catalyst, regenerate to restore activity**
  - Monitor temperature profiles in RGB and first alkylator bed
  - Bypass RGB during catalyst removal for ex-situ regeneration
  - Continue to operate alkylator at full EB production rate

Badger
Licensing LLC

Strengths of EBMax Process

- **Outstanding Commercial Success**
  - 16.2 million metric licensed - 31 plants
- **Process optimized to exploit catalyst strengths**
  - Operation of alkylation at low benzene-to-ethylene feed ratio reduces utilities and investment
  - Low temperature operation reduces steam consumption
- **Bottom line**
  - Nearly stoichiometric yields (low residue)
  - Low catalyst loads
  - Reduced steam consumption
  - Lowest investment and operating cost

Badger
Licensing LLC

Styrene Technology

- **Low steam-to-oil, high selectivity design**
  - 6:1 molar ratio, 97 mole % selectivity
- **Robust and maintainable reheater**
  - Directly underneath reactor (close-coupled)
- **Specialized feed/effluent exchanger**
  - Produces high pressure steam
  - Vertical orientation, occupies minimal plot area, constant pressure drop
- **Catalyst Stabilization Technology provides unique operating flexibility**
- **SM distillation scheme reduces energy use**

Badger
Licensing LLC

Styrene Milestones

1964	Introduction of first generation at Seadrift
1968	First world-scale plant at Amoco Texas City
1973	First Vacuum Dehydro at Foster Grant
1980	Commercialization of fourth generation at Bayport
1986	Introduction of fifth generation at YNCC
1992	Commercialized of PAR Technology at CPChem
1997	Sixth generation startup at SK Oxichemical
1999	Commercialized CST Technology at CPChem
2003	First 6:1 Steam-to-Oil Operation at BASF
2007	Commercialized Multi-Effect Distillation at Samsung
2008	Demonstrated Direct Heating at TOTAL

51 Licensed Plants

Badger
Licensing LLC

Simplified Dehydrogenation Flowsheet

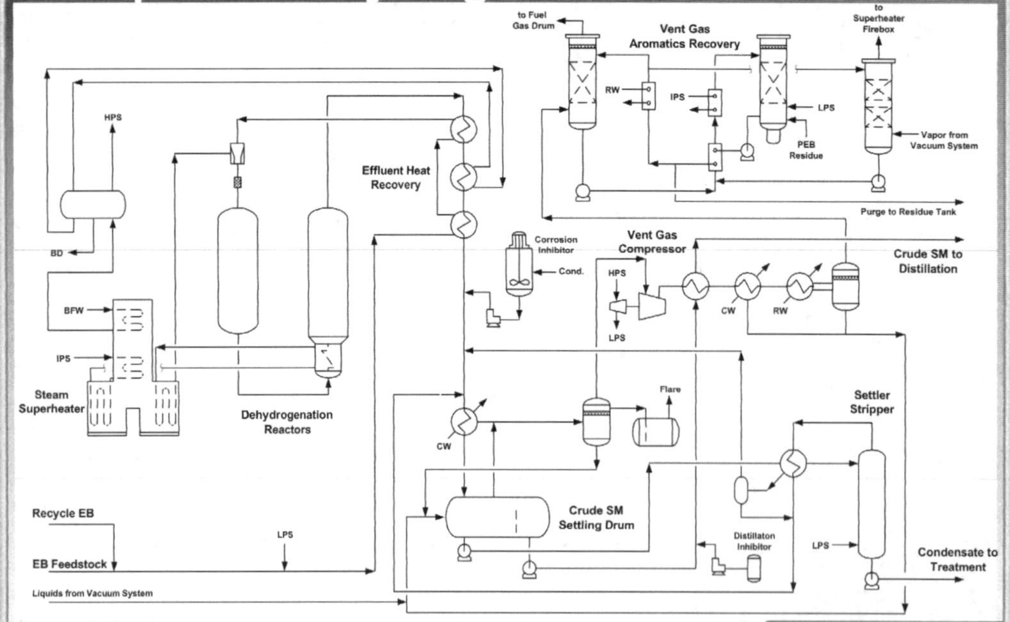

Badger
Licensing LLC

Dehydrogenation Layout

Badger
Licensing LLC

Dehydrogenation Layout at 720,000 MTA

Badger
Licensing LLC

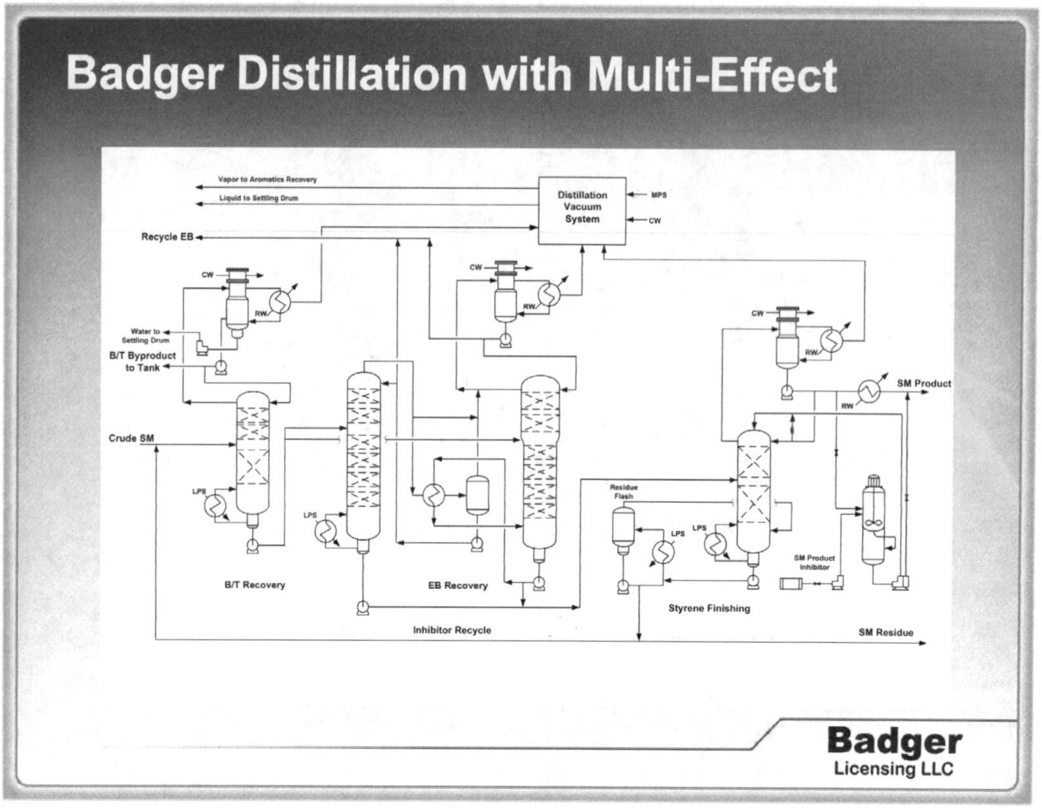
Badger Distillation with Multi-Effect
Vapor to Aromatics Recovery
Liquid to Settling Drum
Distillation Vacuum System
MPS
CW
Recycle EB
RW
Water to Settling Drum
B/T Byproduct to Tank
Crude SM
LPS
SM Product
Residue Flash
SM Product Inhibitor
B/T Recovery
EB Recovery
Styrene Finishing
Inhibitor Recycle
SM Residue
Badger
Licensing LLC

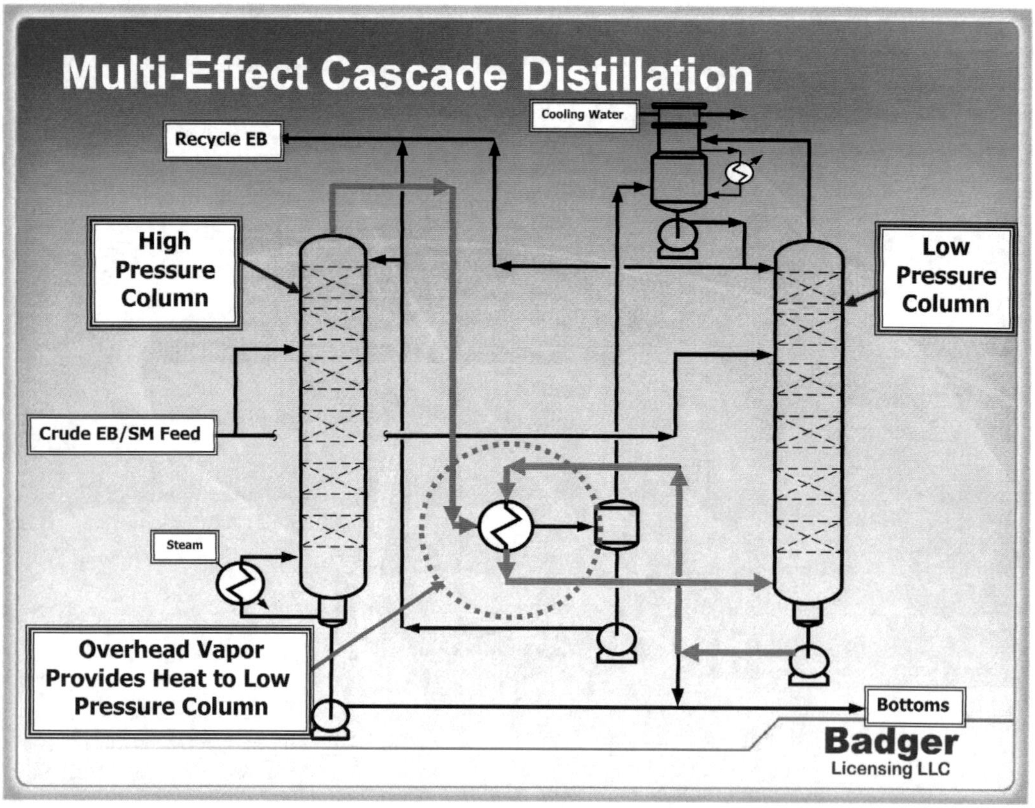
Multi-Effect Cascade Distillation
Cooling Water
Recycle EB
High Pressure Column
Low Pressure Column
Crude EB/SM Feed
Steam
Overhead Vapor Provides Heat to Low Pressure Column
Bottoms
Badger
Licensing LLC

Multi-Effect Cascade Distillation

Badger Licensing LLC

Multi-Effect Distillation Advantages

- **Compared to single-column system**
  - Steam consumption reduced by 45 to 47%
- **Saves about 240 to 285 kcal/kg SM**
  - Steam/CW savings worth about 9 to 10 $/MT
- **Design does not increase polymerization inhibitor usage (as compared to single column design)**
  - Only half the EB/SM separation occurs at elevated pressure thus high bottoms temperature (less than120°C)
  - Low pressure column polymer make extremely low with bottoms temperature between 82 to 85°C

Badger Licensing LLC

Dehydrogenation Catalyst Chemistry

- **Activity depends on potassium in catalyst**
  - A catalyst promoter
  - Catalyzes decoking
- **However, potassium is mobile**
  - Potassium migrates out of the pellets and
  - Migrates within the catalyst pellet
- **As potassium leaves the active sites,**
  - The catalyst deactivates and eventually, needs replacement

Badger
Licensing LLC

Catalyst Stabilization (CST) Concept

- **CST adds potassium to the reactor feed**
  - Low dosage
- **Aged catalyst can be rejuvenated**
  - Restores potassium to active sites
- **Catalyst can maintain recovered activity**
  - Potassium vapor/solid equilibrium affected
- **Higher activity improves run average performance**
- **Concept is covered by Badger and TOTAL patents**

Badger
Licensing LLC

Dehydrogenation Catalyst Activity - Fresh

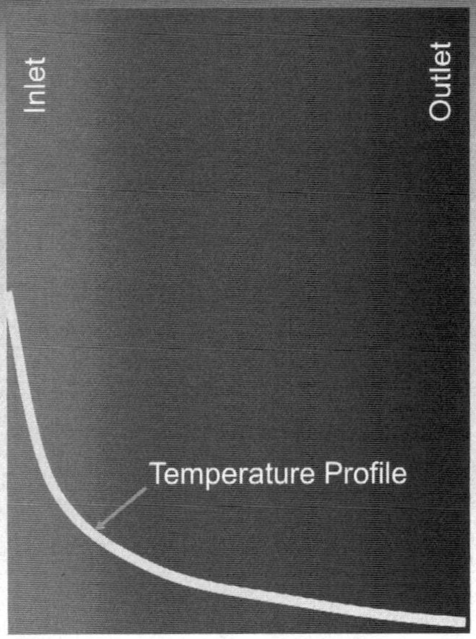

- **Most of the reaction takes place in the first 25% of the bed**
  - Reactors sized based on required run length
  - Could use less catalyst if able to slow down aging

Dehydrogenation Catalyst Activity – Near the End

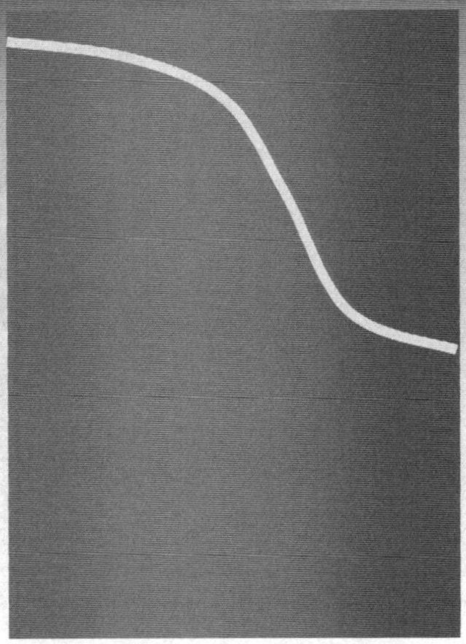

- **At end of run more than 50% of the bed is essentially inactive**
- **The hot void volume expands into the catalyst bed**
  - Benzene, AMS, DVB, heavies all increase
- **The effective aging rate (°C/month) increases**

Badger
Licensing LLC

Dehydrogenation Catalyst Activity – Enter CST

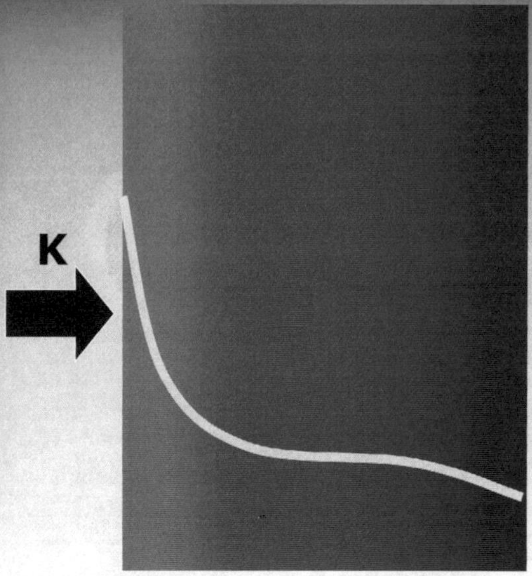

- **The front part of the bed is reactivated**
- **The average bed temperature is reduced**
  - Benzene, AMS, DVB, heavies all decrease
- **Aging is essentially stopped**

Badger
Licensing LLC

CST Commercial Experience

- **Chevron Phillips (American Styrenics) in Louisiana**
  - Two styrene units, each with a CST Unit
  - CST has been used at this site for 11 years
  - 9 catalyst runs with CST
- **Westlake in Sulphur, LA**
  - One catalyst run with CST
- **CST used commercially with 10 catalyst loads**
- **7 additional CST systems licensed in last 3 years**

Badger
Licensing LLC

Overall Catalyst Activity

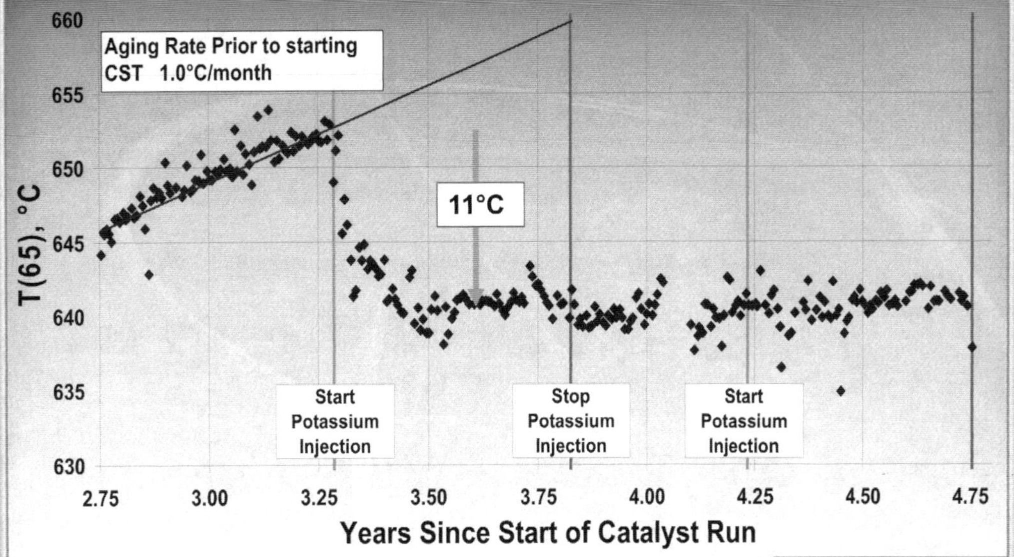

Badger
Licensing LLC

Badger EB/Styrene and Phenolics Technologies

- All core technologies are extremely competitive with leading position in the licensed marketplace
- All technologies have highly competitive economics
- Special attention to practical issues affecting operability of designs
- Extensive research programs to maintain competitive position

Badger
Licensing LLC

Thank You

Questions?

Badger
Licensing LLC

AIChE

Role of the Engineering Community to Address Energy and Sustainability Issues

June C. Wispelwey
Executive Director, AIChE

AIChE Regional Process Technology Conference

- Fulfilling the mission
- Key issues
- AIChE's response
- Trans-disciplinary cooperation

October 2, 2009

AIChE

Our Strategic Goals

- Become a ***global*** organization of chemical engineering practitioners
- *Strengthen **industry and technology groups** and create new groups where needed to support the diverse interests of members*
- Aggressively develop innovative new products and services for members based ***via web-based technologies***
- Engage with others to improve the undergraduate curriculum in chemical engineering and promote ***life-long learning***
- *Impact **societal issues** by informing and educating the public and government in complex technical areas*

October 2, 2009

AIChE

Initiatives

Energy

- 2005: Presidential Commission headed by John Chen
- 2009: 20 activities focused on
  - Programming
  - Member communication
  - Outreach
  - Projects to incubate new programs, communications, outreach and knowledge
- Forming a Center for Energy Initiatives

Institute for Sustainability

- 2004: Presidential Study
- 2005: Technological society (IfS) launch
- 2009: Highlights
  - 1000 professionals
  - Industry consortium establishing best practices
  - AIChE Sustainability Index (a benchmarking service)
  - Progressing the measurement of sustainability with full cost accounting

October 2, 2009

AIChE

The Issue

- Energy supply will not meet demand in the long term
- Use of energy has environmental and societal impacts which has raised public alarm (e.g., greenhouse gases)
- Resulting global legislation and policy decisions is affecting all aspects of our economy and energy security
- Technology is viewed as both "problem" and "solution"
- Selection of energy choices and needed development requires consideration of the "sustainability criteria suite"
  - Technical feasibility
  - Economics
  - Environment
  - Societal impacts

October 2, 2009

AIChE

U.S. Energy Supply

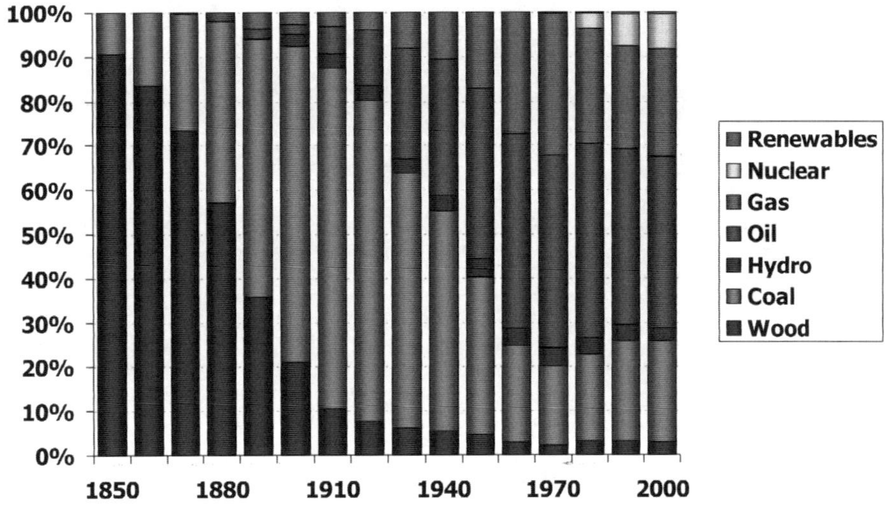

Fossil Fuels Have Ruled

October 2, 2009

Author: Koonin Source: EIA

AIChE

Fossil Fuel Reserves

	Recoverable Reserves, GTC	Reserve Life @Current Rate, Yr	Reserve Life @Projected GDP Growth, Yr
Oil	120	35	25
Natural Gas	75	60	45
Coal	925	400	100

Supply is not sustainable

October 2, 2009

Source: EIA

AIChE

Electricity Usage

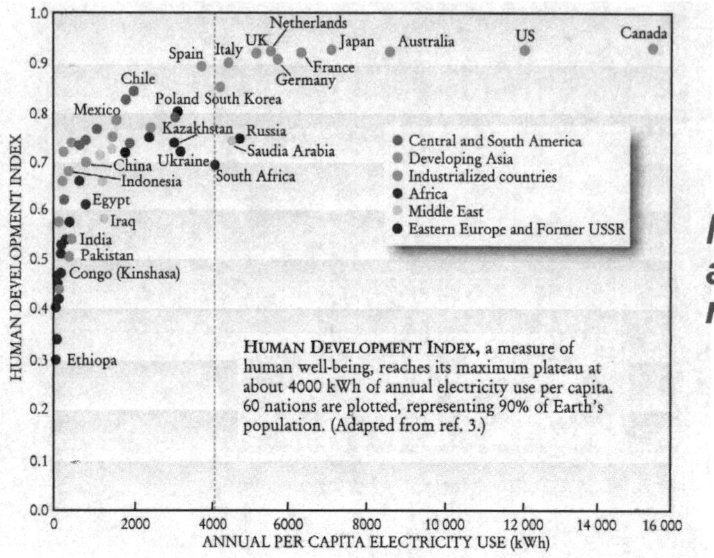

More issues as demand ramps up

October 2, 2009

Source: UN

AIChE

Minimal Short-Term Impact of Renewables

Total primary energy: 410 EJ/year **Renewable: 4 EJ/year**

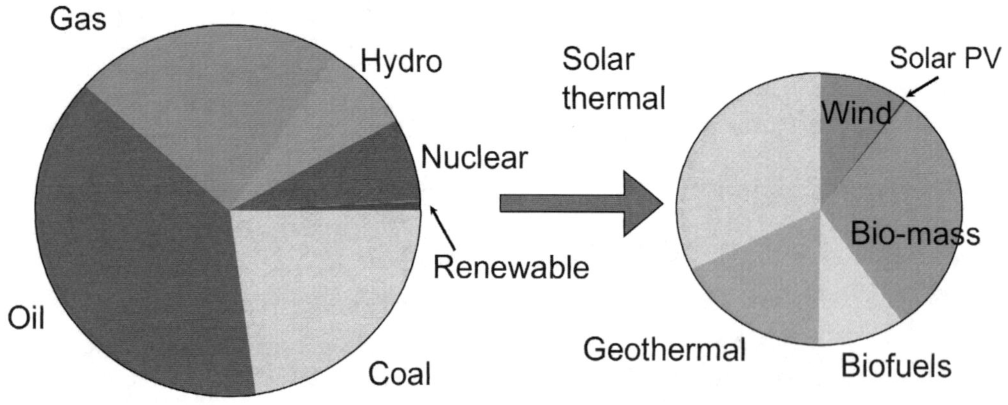

October 2, 2009

Source: EIA

AIChE Greenhouse Gases

Linkage with energy *Public alarm raised*

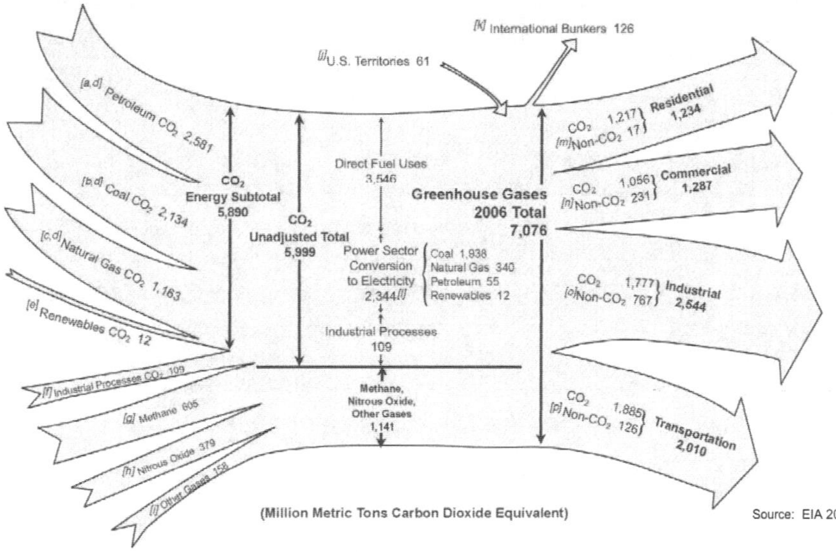

Source: EIA 2006

AIChE Increasing level of Environmental Regulations

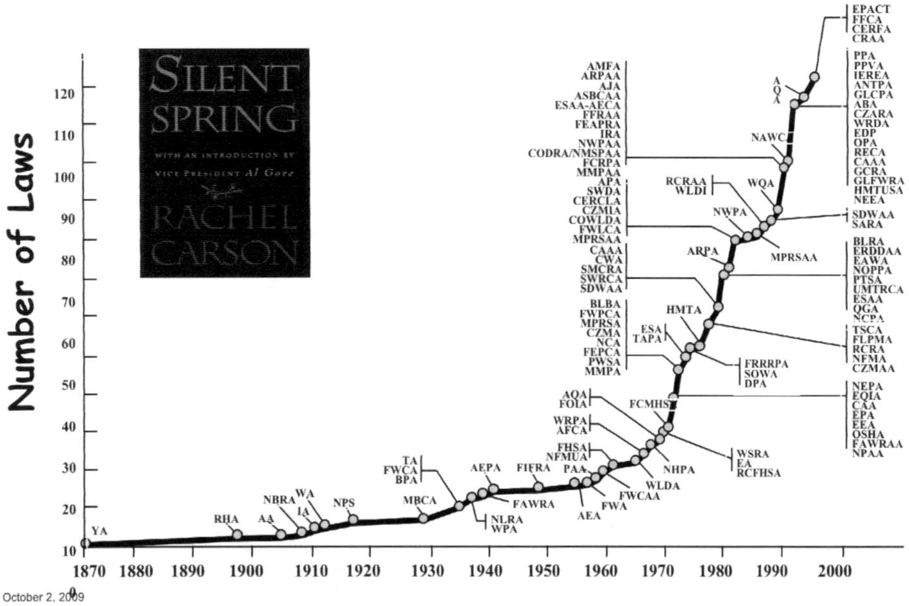

AIChE

CO_2 Targets Could be Huge

IPCC* 4th Assessment Report

Temperature Increase (°C)	All GHG (ppm CO_2 eq.)	CO_2 (ppm)	CO_2 emissions 2050 (% of 2000 emissions)
2.0-2.4	445-490	350-400	-85 to -50
2.4-2.8	490-535	400-440	-60 to -30
2.8-3.2	535-590	440-485	-30 to +5
3.2-4.0	590-710	485-570	+10 to +60
4.0-7.0			+135 (Reference)

Projects suggest that climate might change at current CO_2 production rates, scaring public.

October 2, 2009

* IPPC: Intergovernmental Panel on Climate Change

11

AIChE

U.S. Government Investments in R&D*

To date, too little investment

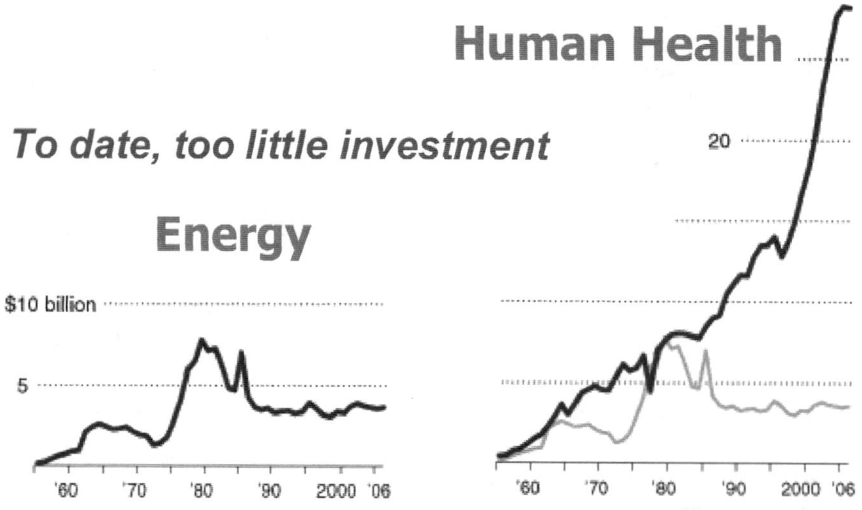

October 2, 2009

*Sources: AAAS 2006; NYT, October 30, 2006. 2006 dollars

AIChE Public Acceptance

Technology seen as "Problem" And "Solution"

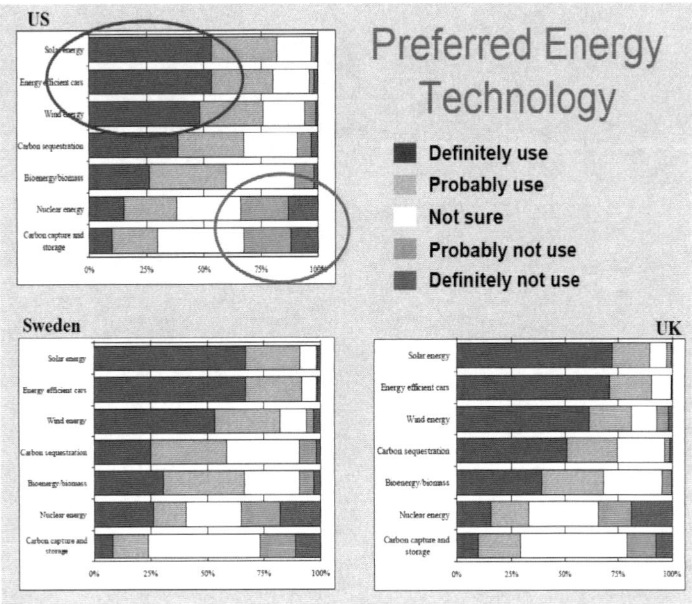

October 2, 2009

AIChE AIChE's Energy Initiatives

- Programming
- Member Communications
- Outreach
- Projects

October 2, 2009

AIChE Programming

- National and Annual Meeting energy sessions
- Energy program guide: National and Annual Meetings
- Specialty conferences
  - Institute for Sustainability (IfS – August 2009 in Cincinnati, OH)
  - AIChE & the Society for Petroleum Engineers' Strategies for Managing CO_2 (Feb 22-24, 2009)
- Participation in other society conferences
  - ASME
  - IEEE
  - international conferences – Mexican and Canadian societies

October 2, 2009

AIChE Member Communication

- Energy website (www.aiche.org/energy)
- Energy newsletter
- Speaker's Academy and Speaker's Corner
- "ChemE on Demand" - *new*
  - An e-learning initiative
  - Multimedia digital library
  - Training center
  - Information forum
  - ***http://apps.aiche.org/chemeondemand/home.aspx***
- Critical issue series in AIChE's magazine *Chemical Engineering Progress (CEP)*

October 2, 2009

AIChE Outreach

- Energy Website
- Energy Newsletter
- ChemE on Demand
- Position and policy papers
  - Massive electric energy storage
  - ACS/AIChE energy position paper
- Congressional briefings

October 2, 2009

AIChE

Projects are Transdisciplinary

- Founder Society's Technologies for Carbon Management
- Renewable transportation fuels metrics
- Education projects for undergraduates
  - Connecting undergraduates to K-12
  - Award for solving energy problems
- Certification
  - Center for Sustainable Technology Practices' (CSTP) industry roundtable on carbon management certification
- Washington Internships for Students of Engineering (WISE) program
- Engineers Without Borders

October 2, 2009

AIChE

More Participation

Premise: U.S. engineers should be participating *more fully* in the "sustainability" and "energy" agenda, and the five founders' societies, with one million members, should be able to help.

Engineers *are* participating, as countless new environmentally responsive technologies are being developed. So, what does "more fully" mean?

An expanded view of the profession that encourages:

- education in environmental science
- increased involvement in public policy
- comfort with environmental values
- going global (seeing the planet as the system boundary)

Today, only a handful of technically trained people are on Congressional staffs and in the major environmental NGO's.

Deeper and more effective impact on policy → Career tracks in engineering and policy → The creation by the societies of new opportunities for their members.

October 2, 2009

AIChE

Implementing Technologies for Carbon Management – Phase I

Development of the Technical Status of Carbon Management in the U.S.

Objective to more actively engage society members in the energy challenge and participate to inform public policy

Founders Society Project

AIChE, AIME, ASCE, ASME, IEEE

Funded by: United Engineering Foundation

October 2, 2009

AIChE

Focus on Power and Transportation

- Selected electric power and transportation segments (>60% of U.S. CO_2 emissions)
- Scope
  - Identify current and future technology options
  - Develop evaluation methodology, ie boundaries and metrics
  - Identify existing study or technology gaps
  - Determine barriers, ie cost, acceptance, performance, or data to carry out analyses

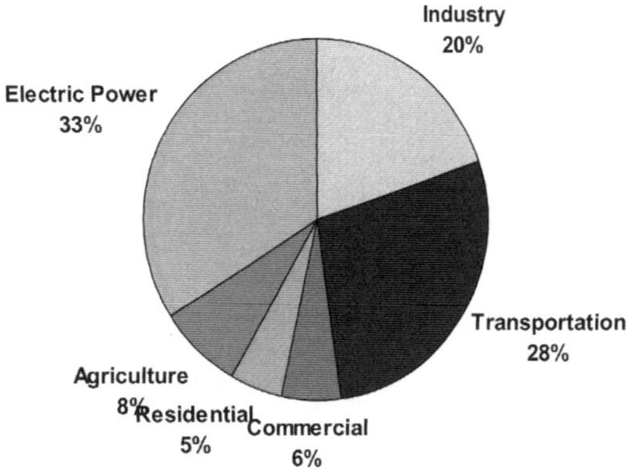

CO_2 Emissions - 2006

October 2, 2009

The Technological Challenges

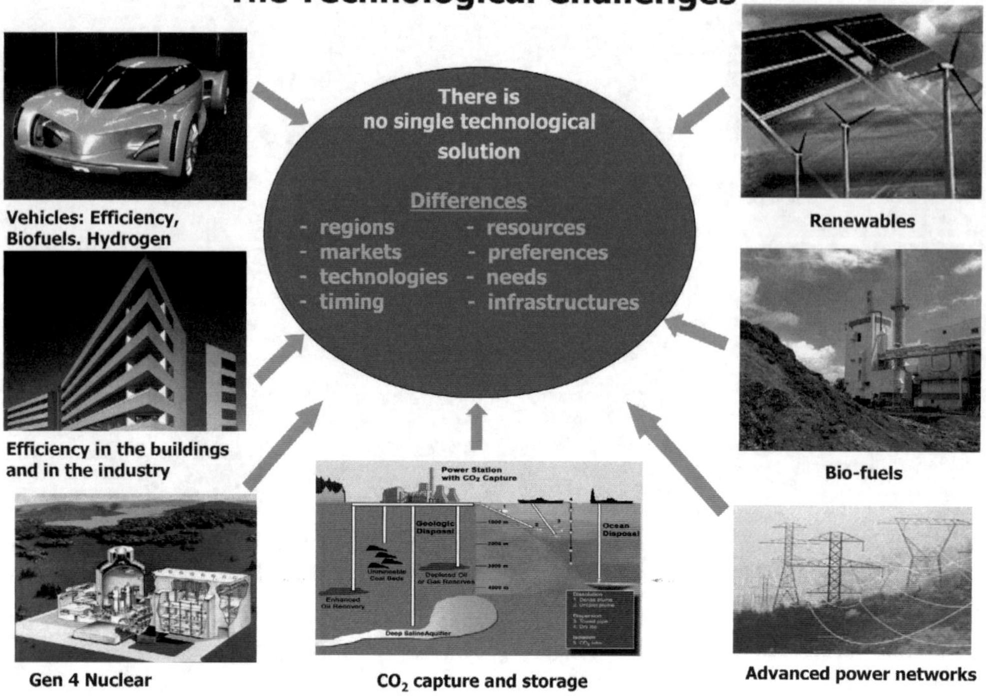

October 2, 2009

AIChE Status of Energy Technologies

Power Generation Technologies

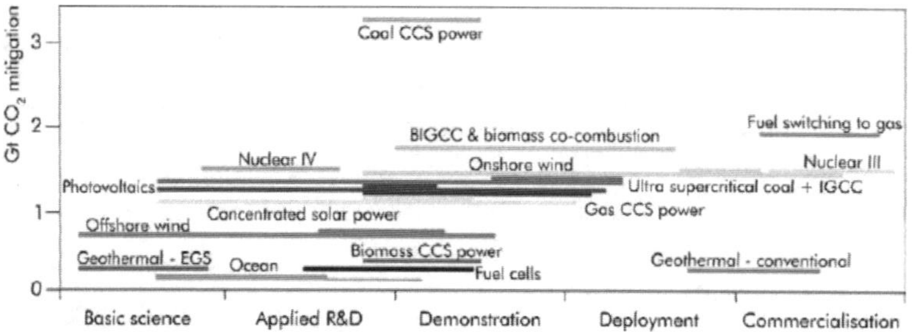

Industrial Technologies

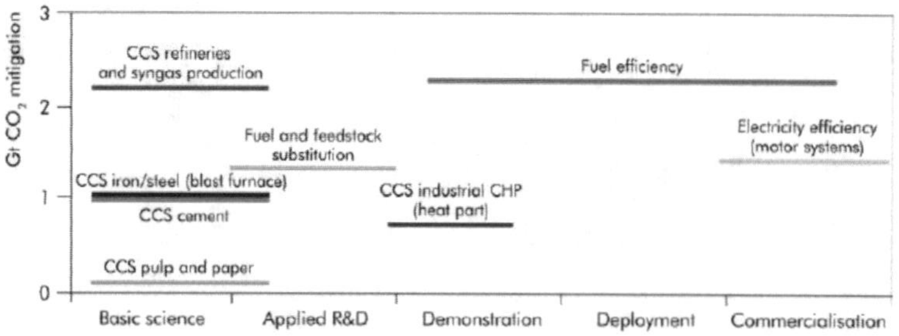

October 2, 2009

AIChE A Transdisciplinary Approach

- Provide a basis for dialogue among the technical community, the public, and public policy makers
- Develop a framework for guiding technology assessment and public policy planning and analysis
  - System boundaries
  - Metrics for evaluation
- Utilize available energy system studies
- Identify gaps and barriers
- Prepare summary documents

http://www.aiche.org/FSCarbonMgmt/

October 2, 2009

AIChE Conventional Electric Power Boundaries

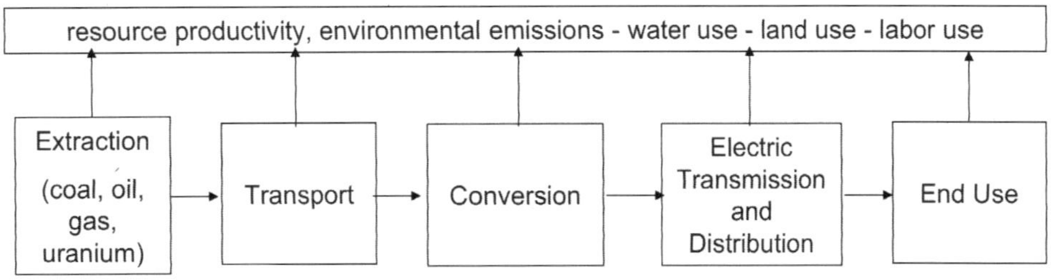

Typical project performance and cost limited to conversion

Renewable electric power has similar supply chain

October 2, 2009

AIChE

Potential Metrics

- Environmental security and sustainability
  - Carbon intensity (g CO_2/product unit - g CO_2/kWhr or g CO_2/mile traveled)
  - Land intensity (unit product/acre - e.g. gal fuel/acre)
  - Water intensity (water use/unit product)
  - Non-CO_2 emissions (e.g. air, water, solids)
- Stable and sustainable energy supply
  - New energy ratio (delivered energy/life-cycle energy input - note some use life-cycle non-renewable energy input)
- National security
  - Balance of trade (e.g. import value/unit product; technology export/import)
- Affordable
  - Life-cycle cost (delivered energy cost/product unit)

October 2, 2009

AIChE

Gaps

- A work in progress with an upcoming workshop (Baltimore, Oct 09)

- Two types of gaps
  - Gaps in existing studies
    - e.g., a given well-to-wheels study may have included economic and energy balance analysis but excluded land intensity metrics
  - Technology gaps limit implementation of a given technology
    - e.g., understanding the geologic CO2 storage capacity and potential leakage for a given location

October 2, 2009

AIChE Barriers

- Data to permit comprehensive system analyses
- Resources (e.g. R&D funding trained work force)
- Time to realize terawatt capacity (this includes time for scale-up, requirement for technical breakthroughs, etc.)
- Public acceptance / understanding
- Competitive cost
- Regulatory framework enabling implementation (includes financing options)
- Performance (e.g. reliability)
- Integrated system implementation (need for availability of all supply chain components)

October 2, 2009

AIChE Next Steps

- Focus on scenarios, not a roadmap (we do not have a destination)
- Boundaries need to include integrated energy system
  - not limited to components or sub-systems
- Focus on how to overcome a given energy system's constraints
- Gaps and barriers workshop (10/09)
  - Near-term improvements: efficiency, energy productivity, replacement technology, CO_2 capture and sequestration
  - Implementation barriers
  - Data gaps for life cycle analysis
  - R&D gaps
- Metrics
  - % of segment that is carbon free (establishes benchmark; identifies needs; and becomes a score card)
  - Qualification of technologies - readiness, affordability, public acceptance

October 2, 2009

Website: http://www.aiche.org/FSCarbonMgmt/

AIChE

5 Ways to Cut Emissions and Energy Usage

	1 Ton Carbon/yr	Cut in Half	How?
Drive	10,000 mi, 30 mpg	60 mpg	Lighter cars using less fuel/power
Drive	10,000 mi, 30 mpg	5,000 miles	Live closer to work
Fly	10,000 miles	5,000 miles	Video-conference
Heating Home	Nat. gas, avg. house, avg. climate	Insulate, double-pane windows, fewer leaks, condensing furnace	
Lights	300 kWh/month when all-coal power (600 kWh/month, NJ)	Add CCS* at 60% of the plants, **or** Permanently replace twenty 60W incandescent bulbs, lit 6 hrs/day, with compact fluorescents.	

Source: Robert Socolow, Princeton

October 2, 2009

*CCS = CO_2 capture and sequestration

Distillation Device Technology.

A Mature Field?

Jose L. Bravo
Shell Global Solutions

The fundamentals..What can the distillation device do

- Colburn Analogy
  - $f/2 = k_x/(\rho v)^* (\mu/\rho D)^{2/3} = h/(cp/\rho v)^*(c_p\mu/k)^{2/3}$
  - *Film drag* α *Mass transfer coefficient* α *Heat transfer coefficient*
  - *(form drag is not very useful)*

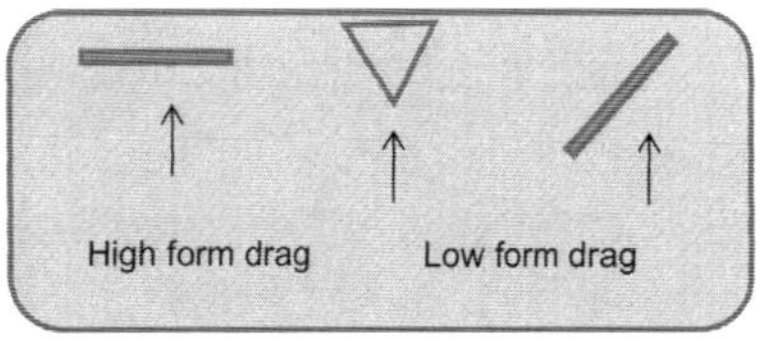

- Eckert Correlation
  - *Capacity limited by disengagement*
  - *Disengagement propension* α *gravity and 1/dp*
- Two-Film/Penetration Theory
  - *Mass transfer flux* α *interfacial area and renewal rate*

1

Eckert Correlation...... Hydraulic Capacity

Confidential

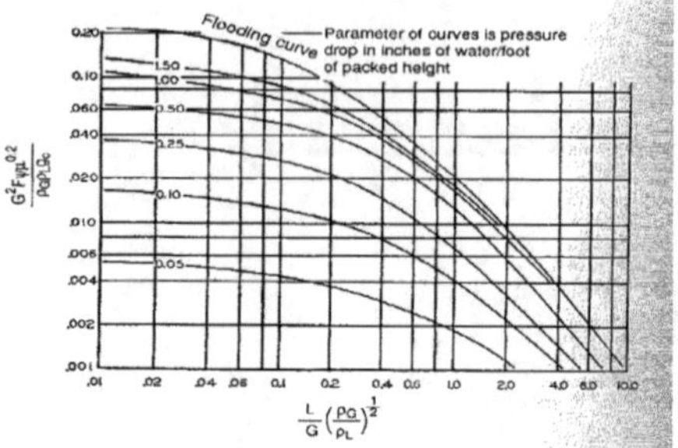

L = liquid rate, lb/s, ft²
G = gas rate, lb/s, ft²
ρ_L = liquid density, lb/ft³
ρ_G = gas density, lb/ft³
F = packing factor
μ = viscosity of liquid, centipoise (cP)
ψ = ratio, $\frac{\text{Density of water}}{\text{Density of liquid}}$
g_c = gravitational constant = 32.2

Figure 8.17 The Eckert 1970 version of the GPDC correlation. (From J. S. Eckert, *Chem. Eng. Progr.*, *66*(3), March 1970, p. 39. Reproduced courtesy of the American Institute of Chemical Engineers.)

- Hydraulic capacity depend on the ability for the phases to disengage after contacting
- Stokes' Law analogy.
- It is the effective density difference of the phases so higher g values increase separability
- Higher g values can be obtained depending on device design

2

So what have we seen over the years?

- Trays
  - o Intimate phase contact (Bubble cap Trays)
  - o Lower form drag (sieve and valve trays)
  - o Lower hydrostatic losses (MD, HiFi trays)
  - o Enhance gravity (ConSep, Ultrafrac)

time

Efficiency has more to do with the system than the device if the latter is applied smartly. . . (O'Connell)

- Packings
  - o Lots of surface and mixing (Berl Saddles, Raschig Rings)
  - o Reduce form drag (See Colburn) (Pall Rings, IMTP
  - o Reduce form drag and increase surface (Structured Packings)
  - o Reduce form drag some more (Newer structured packings)

Reliability, retrofit-ability, good phase distribution, and low installed cost as parameters

3

So really

- What we do:
  - o Reduce form drag
  - o Increase interfacial area
  - o Increase gravity
  - o Try to keep liquid distribution at an optimum
- We apply also to
  - o inlet devices
  - o phase separators
  - o distributors, etc

The question, How can we change the game?

4

Process Intensification in Distillation Column Internals and Gas/Liquid Phase Separators .

- Separation between phases of different densities is the controlling factor in the capacity of phase separators, distillation towers, absorbers, strippers, etc
- Separation of phases is enhanced in "high gravity" environments
- g's need to be generated to intensify the process
- Simplicity!!!!!!
- Ability to retrofit to existing assets

5

Capacity of Phase Separation Devices

- **Impingement Separators**
  - Mesh mist eliminator
    - Maximum load factor 0.1 to 0.2 m/s
  - Vane mist eliminator
    - Maximum load factor 0.15 to 0.25 m/s
- **Swirl Tube™ mist eliminator**
  - Maximum load factor (tube) 0.4 to 0.8 m/s

$$\textbf{Load factor} = U_g\,(\rho_g/\rho_l-\rho_g)^{1/2}$$

6

The Shell Schoepentoeter™ Inlet Device

- Lowest pressure drop device for two phase inlets.
- Good vapor distribution capability
- Allows for use of smaller inlet nozzles for reboiler returns and flashing feeds. No need to enlarge nozzles when debottlenecking.
- Good de-entrainment characteristics
- Requires minimum column height of any inlet device/vapor distributor.

7

Shell Schoepentoeter™ Inlet Device

8

Schematic of a Separation Swirl Tube

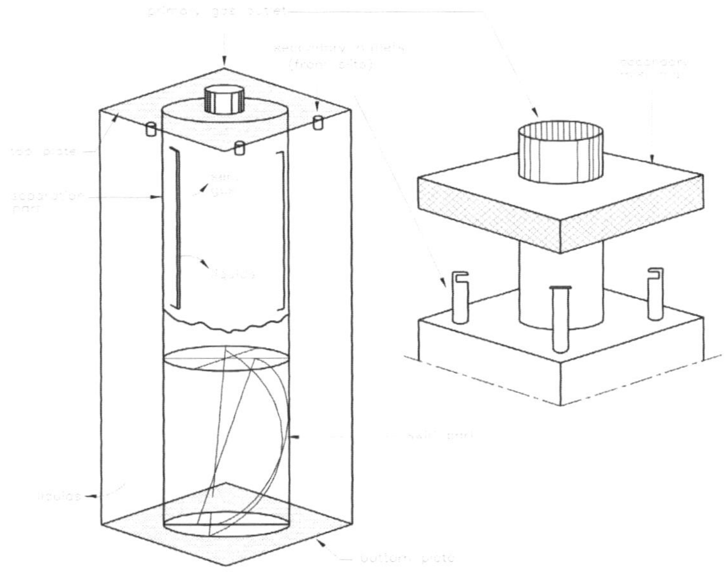

9

Shell SMS™ Separators

- Integrated phase separation equipment that includes Schoepentoeter, Mist mat, and Swirl Deck technology for the most efficient droplet removal.
- Very high capacities and efficiencies
- Used in new and debottleneck applications

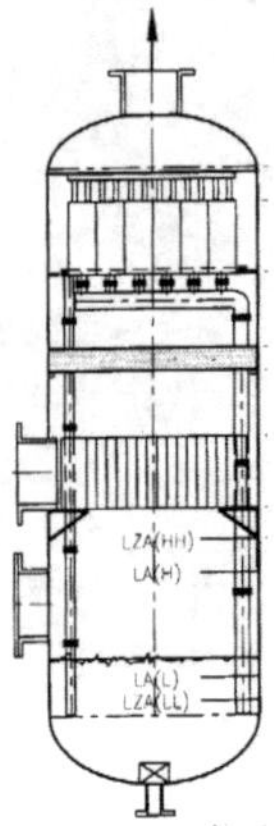

10

Shell ConSep™ tray

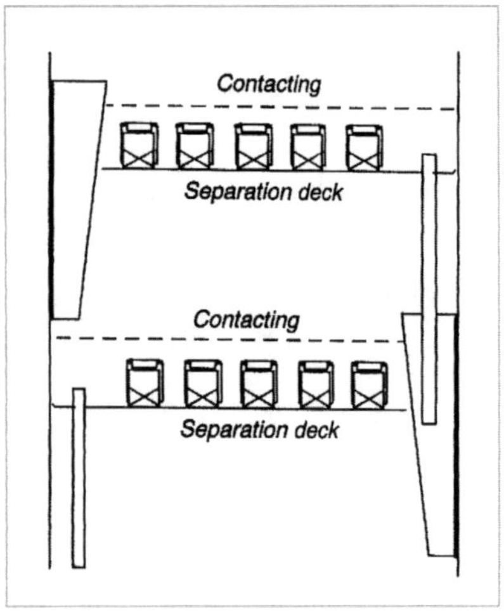

- Combines features of contacting tray and centrifugal separator
- Allows operation above floodpoint
- Tray efficiency and tray spacing comparable to conventional trays

  450 mm < TS < 800 mm
- Inexpensive debottlenecking option for existing columns

Nibble around the edges

- Surface area increases.- Would need order of magnitude change.... Tough
- Increase void fraction.- not likely
- Reduce form drag.- many advances.... What is left?
- Increased gravitational fields...Room here for more g's, smaller drops, better disengagement... Current solutions are +150%.....Too much?
- Minimize weir loadings while maximizing bubbling area....Incremental improvements may be available

- But: What we are thinking about today is incremental, and emphasizing survival in lower margins, All signs of maturity... But we have been here before (40 years ago) and then boom! Structured packings appeared

12

Distillation Device Technology is Mature! (But it does not have to)

- Yes, Why?
  - o We haven't come up with something new in the recent past. Just marginal improvements I say
  - o So yes it is mature but it is our own fault!
  - o It is up to us to make it immature!!!!!! By innovation and discovery

- Do I have the answer? Of course not. But how about looking someplace else besides Colburn, Eckert?
- For example: Surface renewal at the vapor/liquid interface. How can a device really improve renewal in both phases and in films **and** droplets?
- I believe we are not done playing with gravity
- How about the Continuity Equation? (exchange pressure for velocity with recovery in boiling systems?)

13

One final Point

- Technology is a three-legged stool. You need all three or you tumble:
  - o Good fundamentals
    - ➢ In this case creative devices
  - o A capability to deploy it, understand it, and operate it
    - ➢ Users and Designers need to be knowledgeable of fundamentals, effective diagnostics, troubleshooting, etc
  - o Reliability, cost, and robustness fit for purpose
    - ➢ "A good idea that is too expensive is not a good idea"

14

The Importance of Laboratory and Pilot Testing for Separation Processes

(or Why would you want "Real" data when you can use a computer model?)

Andrew (Jack) Eckles
National Thermodynamics Laboratory, Galveston, TX
www.ntl.us.com

Typical Categories

- Scale up from Laboratory Developments
- Unknown Properties: Pure, interaction, azeotropes, VLE, solubility, etc
- Unknown Side Reactions

Purposes

- DATA FOR ACCURATE MODELING AND SCALEUP
- Data for Optimization
- Safety
- Purities
- Efficiency (conversion, side reactions, etc.)
- Proof of concept

INTRODUCTION

- The Good, the Bad, and the Ugly

Processes

- Distillation
- Absorption
- Reactions
- Stripping
- Stability Testing

Modeling and Calculations

Great For:

- "What If......."
- Determining Initial Operating Conditions
- Setting Boundaries and Limits
- Well Known Processes & Systems
- Economical Testing of New Processes & Operating Conditions
- Tweaking Existing Processes

Modeling and Calculations

Not Optimum For:

- Very High Purities
- Very Low Concentrations
- Thermal Stability Problems
- Azeotropes
- Groups of "Interesting Compounds"
- Materials with Incomplete Data

Economic Implications

- Sizing
- Purities
- Process Times
- Potential Trace Contaminants
- Unknown Side Reactions
- Accumulation of Trace Contaminants

Stripping Examples

- Dioxane
- Oil/Water
- Ethanol
- Food Grade Butane

Process Combinations

- DOE Ethanol Production
- Molecular Sieves & Distillation

3 Stage Continuous Distillation with Molecular Sieve Drying

High Purity Silane Based Material

Cryogenic Test Distillation

Air Sensitive Dopant
(preassembled)

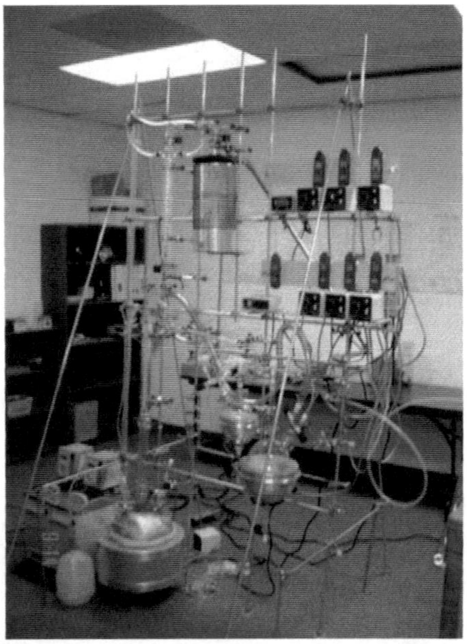

Pyrophoric Dopant Test
30 micron Distillation

High Purity IPA Recycle

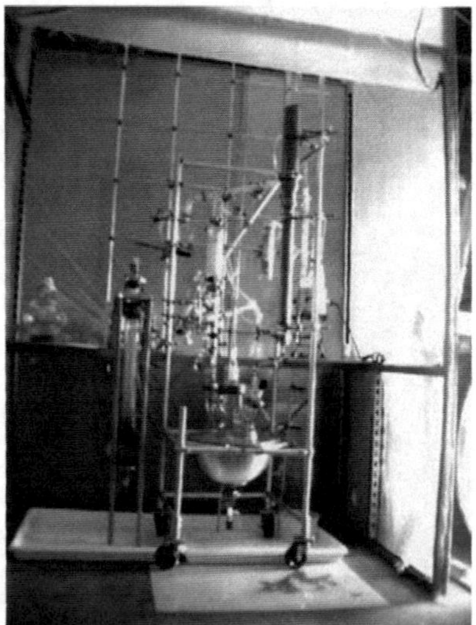

H2O2 With Blast Shield Removed

Pilot Scale CO2 Absorber

Fluorinated Pharmaceutical

Food Grade Butane &Cocoa Butter

Food Grade Butane &Cocoa Butter

Adhesive Synthesis

Pharmaceutical Plant Extracts

Corning Bubblecap

Corning Bubblecap

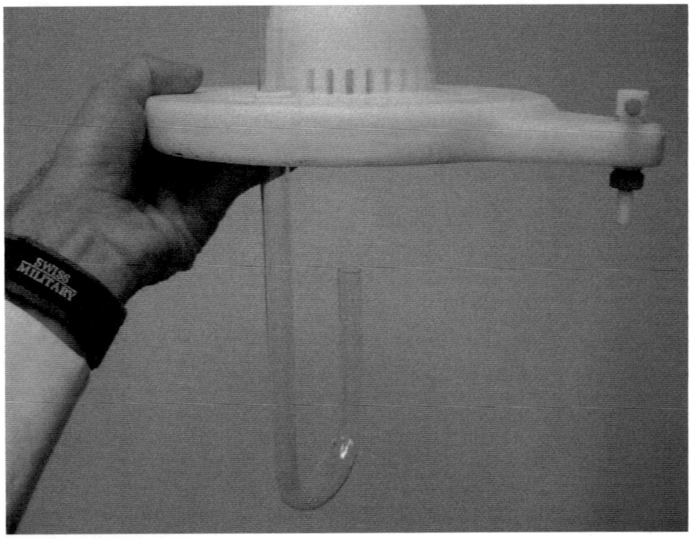

Los Alamos Nitric Acid Recycle

Modeling of Thermodynamics

And Kinetics Has Its Advantages

Troubleshooting the Diameter Transition Zone of a Depropanizer Tower

Pless, L., Tracerco
McKeown, C., Shaw E&C
Xu, S., Shaw E&C
Originally presented on: 10/2/2009 11:00:00 - 11:30:00

A depropanizer tower with a large diameter top section and smaller diameter bottom section was field tested and studied for its root cause of flooding. In the diameter transition zone, a two-phase feed was being introduced. Gamma scans identified that the trays in the top section were flooding and the liquid accumulation initiated in the transition zone. A neutron backscatter technique was used to investigate downcomer hydraulics and identify the root-cause of flooding in the top section of the column.

New Developments in Random Packing of Columns

Nieuwoudt, I., Koch-Glitsch
Quotson, P., Koch-Glitsch, LP
Originally presented on: 10/2/2009 11:30:00 - 12:00:00

Advances in random packing of columns provides improved capacity and efficiency compared to other random packing. Data over a wide range of operating conditions will be presented for several sizes of the new random packing types versus other random packing that are commonly used in industry. Case studies will be used to demonstrate the benefits of this new random packing.

Energy Real-Time Optimizer
Baytown Refining/Chemical Complex

ExxonMobil
Taking on the world's toughest energy challenges ™

Frank Roberto for Maurice Lorenz

AIChE Regional Process Technology Conference
Galveston
October 2, 2009

Generic Refinery Utilities System

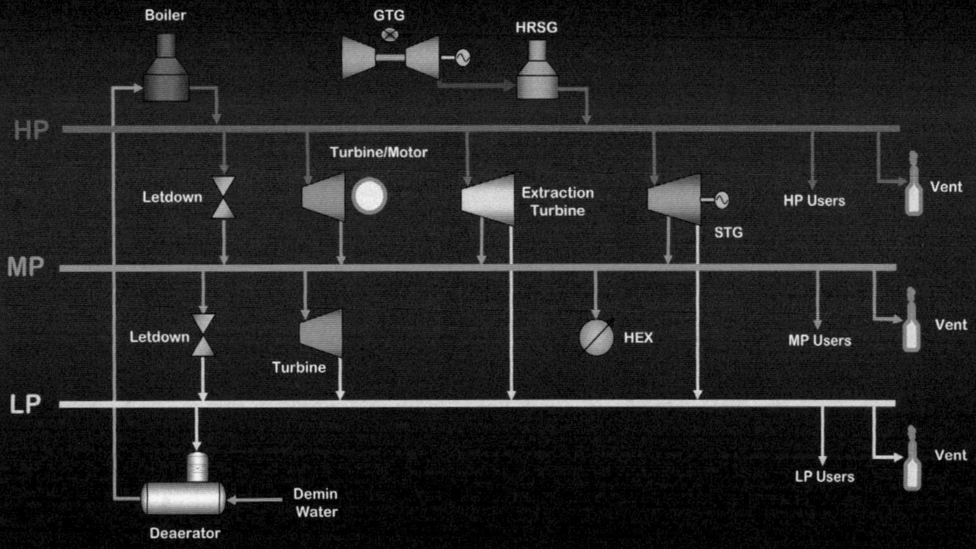

ExxonMobil

-2-

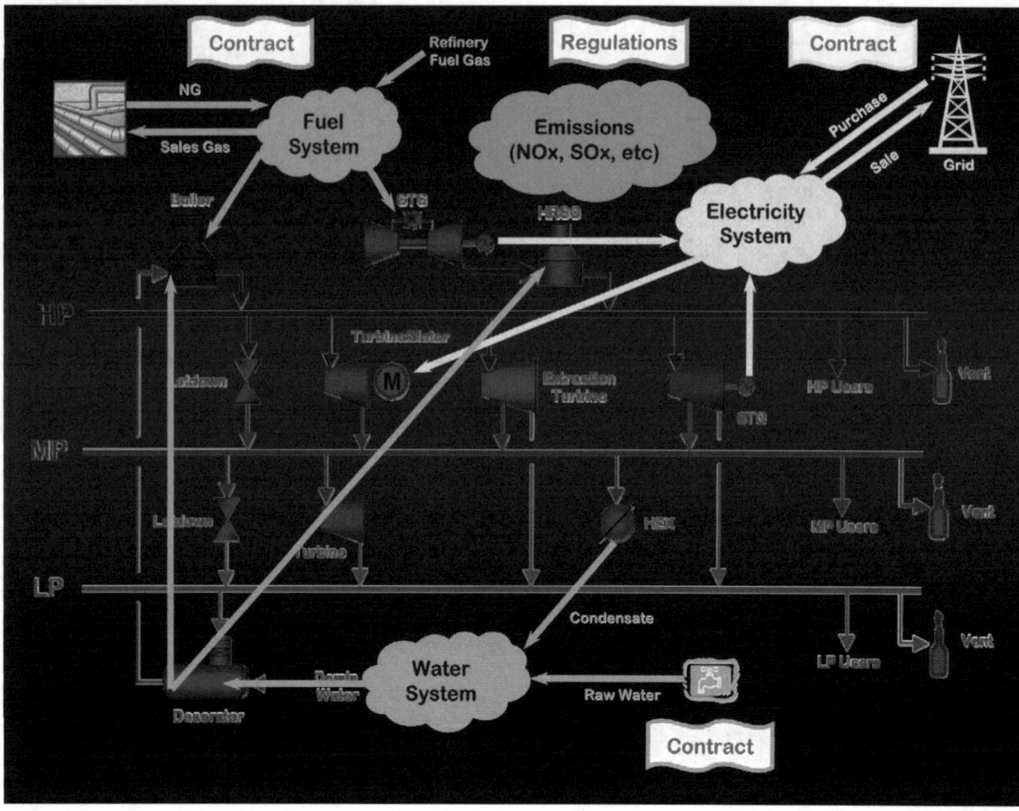

Contract
Refinery Fuel Gas
Regulations
Contract
NG
Sales Gas
Fuel System
Emissions (NOx, SOx, etc)
Purchase
Sale
Grid
Boiler
GTG
HRSG
Electricity System
HP
Turbine/Motor
M
Letdown
Extraction Turbine
GTG
HP Users
Vent
MP
HEX
Turbine
MP Users
Vent
LP
Condensate
LP Users
Vent
Demin Water
Water System
Raw Water
Deaerator
Contract

theLamp
Fall 2003 Vol.85 No.2
With some $5 billion a year of energy required to run ExxonMobil's operations around the world, including the Baytown, Texas, refinery, power-supply management is a major activity.
"It takes $15 million of energy per day or $5 billion of energy per year to run our operations," says Philip Dingle, president of ExxonMobil Gas & Power Marketing Company. "That's a major share of our operating costs. So we have a big incentive to wisely manage energy consumption and costs, not just in refineries, but also in our chemical plants, production operations, marketing terminals, office buildings and other facilities."

Integrated Energy Systems Optimization

- Optimizes mix of discretionary fuels to minimize cost of steam, fuel, & electricity subject to environmental and operating constraints
- Incorporates emissions constraints (NOx)
- Monitors steam, fuel, power usage as well as emissions on a site-wide or single unit basis
- No modification to energy needs of process units except for switching pumps and compressors between motor and turbine drivers

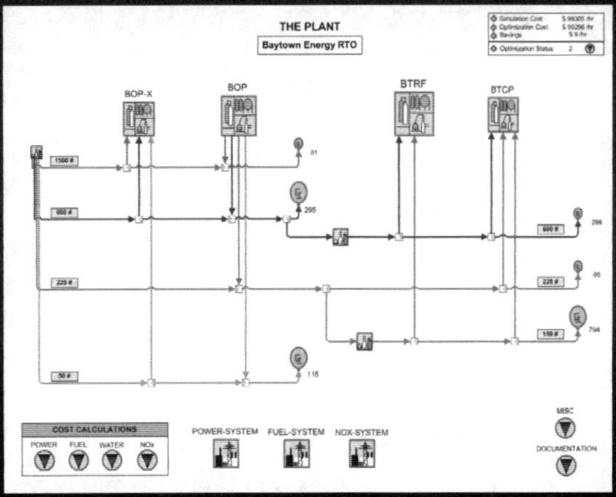

ExxonMobil

- 5 -

Baytown Complex

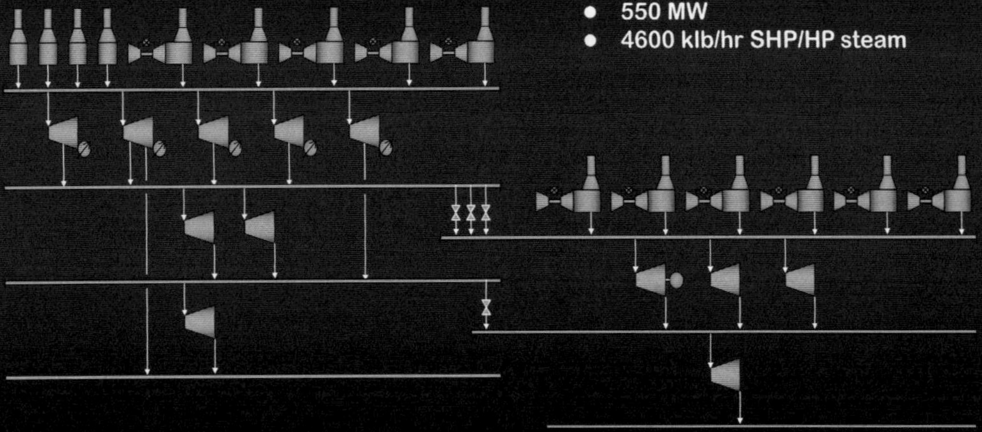

- 11 GTG's with HRSG's
- Supplemental firing
- 4 Conventional boilers
- 550 MW
- 4600 klb/hr SHP/HP steam

ExxonMobil

- 6 -

Baytown Price Volatility

- Power and gas price volatility creates opportunity
  - Power prices changes on a 15 min frequency
  - Ranges from 20 to 200 $/MWh on a typical summer day
  - Gas price changes daily

ExxonMobil

- 7 -

Baytown SPORT Application

- Online Closed-Loop Application
  - Visual MESA™ (Soteica LLC) Software
  - Targets sent directly as set-points to DCS controllers
  - Runs every 15 minutes
  - Spans 3 control centers and 5 consoles
  - Two operating organizations
  - Over 2500 live inputs

- Optimization Handles
  - GTG % Load
  - Steam producer relative load
  - Relative extraction rates for extraction/condensing turbines
  - Steam export rates
  - GTG steam injection
  - Demin water - BFW heat integration
  - Deaerator pressure

ExxonMobil

- 8 -

Baytown SPORT Application

- Constraints Managed
  - Equipment limits
  - Steam cushion
  - Fuel cushion
  - Electric Power distribution
  - Power sales forecast
  - NOx emissions
- Objective
  - Minimize total fuel/power/water operating cost subject to operating constraints

ExxonMobil

- 9 -

Offline Use

- Offline
  - Utilities planning and scheduling
    - Evaluate effect of production plans
    - Evaluate effect of generation equipment maintenance
  - Investment planning and assessment
    - Economics of changes in process steam demand
    - Economics of new cogen
    - Driver selection
    - Evaluate steam system investment

ExxonMobil

- 10 -

Power Optimization

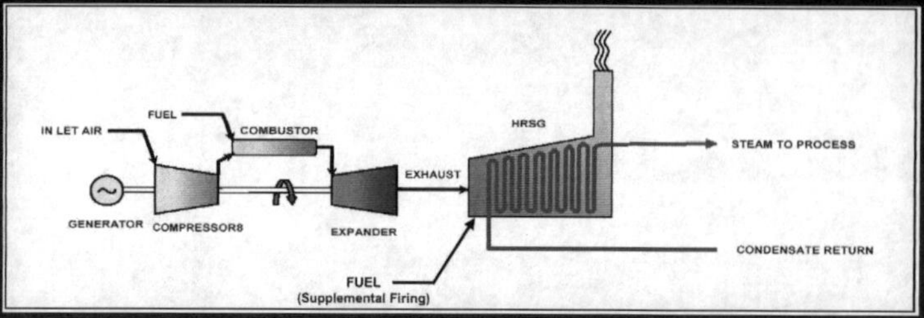

- Power Optimization
  - Base load GTG's vs. turn down and buy additional power
  - Also can vary GTG steam injection
  - Net marginal cost of cogen power is a complex function of:
    - GTG efficiency
    - HRSG Efficiency
    - Alternative steam source efficiency

ExxonMobil

- 11 -

Power Optimization

- SPORT calculates the breakeven point for each GTG
  - If power price below break-even marginal cost of generation, GTG load reduced until break-even point (or min constraint) is reached

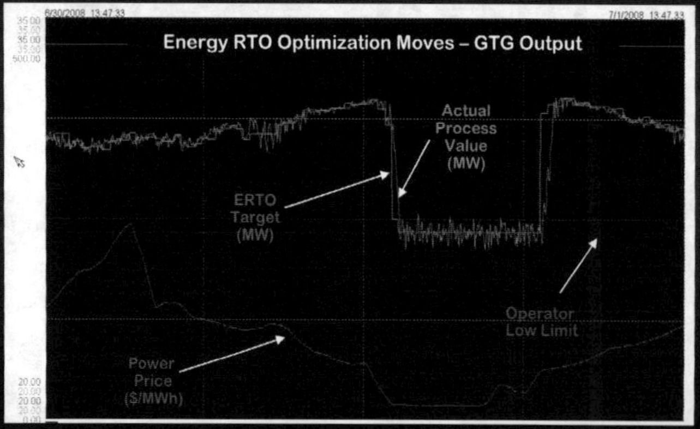

ExxonMobil

- 12 -

Steam Optimization

Several major areas of steam system optimization

- Relative firing of various discretionary steam producers. Cost of each can vary according to:
  - Steam generator efficiency
  - Fuel source
  - Pressure level and mechanism for delivery to common demand point
  - BFW source and deaerator heat source

- Relative extraction rates for extraction-condensing turbines
  - Four parallel process drive turbines deliver 660# steam to complex
  - Each has a different efficiency curve, operating constraints, and condenser pressure
  - Total extraction set by demand, but SPORT optimizes split across four machines

ExxonMobil

- 13 -

Steam Optimization

- Extraction vs. STG load
  - MP steam can be delivered via extraction-condensing turbine or Steam Turbine Generator
  - Extraction steam price a function of fuel only; STG price also a function of power price
  - Thus power/fuel price ratio determines which of these is most attractive

- Deaerator operating pressure
  - Deaerators feed varying mix of boilers, HRSG's and process waste heat boilers
  - BFW temperature has a different effect on the efficiency of each of these
  - When combined with variable source of deaerator heat, this is a complex optimization

- Heat integration of condensate/demineralized water, and low level heat recovery

ExxonMobil

- 14 -

Sustainment and Stewardship

Critical to success of RTO effort:

- Site Operations involvement and buy-in during model development
- Validation and ongoing review of solutions
- Model Owner
- Management of Change
- Track utilization factor
- Keep benefits and Lost Opportunity in front of MGT
- Optimizer failure response procedures
- Training and documentation

ExxonMobil

- 15 -

Benefits

- Operating cost reduction
- Improved constraint management
- Deeper understanding of marginal efficiencies
- Performance monitoring
- Planning tool

Baytown Olefins Plant
Train 5 Cogen Unit

ExxonMobil

- 16 -

Chevron Human Energy | EPCON International | EPI Engineering

Improve Refinery Energy Efficiency with a Cooling Water Fluid Flow Study

Tin Yin Lam*, Todd J. Willman**, Pradnya Bhawalkar**, Arjun Gujar**, Ord Limbrick***
*Chevron, **EPCON International
*** EPI Engineering

Overview of the Fouling Problem

- Bott, who has authored much research on heat exchanger fouling, published a comprehensive study in 1968 and then again in 1990, and remarked that although immense progress has been made in laboratory fouling research, in industry absolutely nothing has changed [1].
- In fact, the only measure currently taken in industry to combat fouling is to oversize heat exchangers to allow for the fouling that will eventually occur anyway.
- "Fouling has been a problem ever since heat exchangers were invented [1]." Heat exchangers are designed for heat transfer. When fouling occurs, a resistance is built up that effectively blocks heat transfer [2].
- Fouling is especially prevalent in cooling water towers and systems. In the top 20 refineries in the United States, approximately 5 – 16 million dollars per year per refinery are spent on cooling water fouling which is approximately 35% of the heat exchanger fouling costs in a refinery.

Savings and Cost Reduction Options

- Temperature should be monitored and controlled carefully with the flow - a temperature difference of just 5 °C can increase fouling by 100% [(1)].
- This can only be reliably done by performing a fluid flow analysis of the entire cooling water system so the system can be quantified and analyzed in detail so that low-cost solutions can be engineered.
- A fluid flow model will allow for testing and verification of restriction devices, piping changes, etc. to balance the flow so that all exchangers are operating close to their design flow.
- The majority of savings are related to the increase in refinery throughput that cooling water fouling is currently limiting.
- Significant, additional savings are also realized in solving root-cause fluid flow problems in cooling water systems.

1. Bott, T.R. Heat Exchanger Fouling and Its Mitigation.

Heat Transfer & Fouling Formulas

$Q = U\ A\ \Delta T_{lm}$

Q = Heat Transfer (Btu/hr)

U = Overall Heat Transfer Coeff (Btu/hr-ft2-F)

A = Exchanger Area (ft2)

ΔT_{lm} = Log Mean Temperature Diff. (F)

$1/U_{fouled} = 1/U_{original} + R_f$

R_f = Fouling Resistance (hr-ft2-F/Btu)

4 Types of Fouling & their Behavior

Fouling can be decreased by increasing the velocity and decreasing the tube wall or metal temperature (7). Uncontrolled water quality can also result in a dramatic rise in fouling (7). The effect of increasing different operating conditions on each type of fouling in cooling water systems is shown below:

Operating Variable Change	**Crystallization & Chemical Reactions**	**Microbial Fouling**	**Particulate Fouling**	**Corrosion Fouling**
Increase in Temperature	mostly increases	mostly increases	mostly increases	mostly increases
Increase in Velocity	decreases/no effect	mostly decreases	decrease	decrease/increase
Increase in pH	increases	increase/decrease	increase/decrease	increase/decrease
Increase in Pressure	increases/no effect	increase/decrease	no effect	increases

Crystallization & Chemical Reactions

- Chemical reaction / crystallization fouling are largely related to cooling water treatment.
- Chemical reaction fouling is present when chemical reactions occur in the fluid and products are deposited on the heat exchanger, such as hydrocarbon cracking and coking (8) (9).
- Crystallization (scaling) occurs when dissolved salts in the fluid precipitate as liquids are heated or cooled.
- Cooling tower water is saturated with air (oxygen). Ideally, the water treatment compounds form a protective layer on the heat transfer surface to protect it from oxidation and electrochemical corrosion mechanisms.
- These compounds (mostly phosphates and their derivatives) are soluble and stable in a narrow range of environments (pH, temperature, concentrations, etc.)
- They undergo chemical transformations and precipitate on a heat transfer surface when the temperature is sufficiently high to promote the reaction.

Other Exchanger Fouling Types

- Microbiological adhesion fouling occurs as macro or micro organisms attach and grow on the heat transfer surface.
- Particulate fouling occurs as suspended solids in the fluid, such as dirt and silt, or particles from the air, such as dust or soot, deposit onto the heat transfer surface (9).
- All types of fouling can lead to corrosion of the material, which is known as corrosion fouling.
- Most often, we have combined fouling, where more than one type is present and the problem becomes very complex.
- By understanding the mechanism by which fouling occurs, we can begin to overcome it.

Fouling Rate vs. Temperature

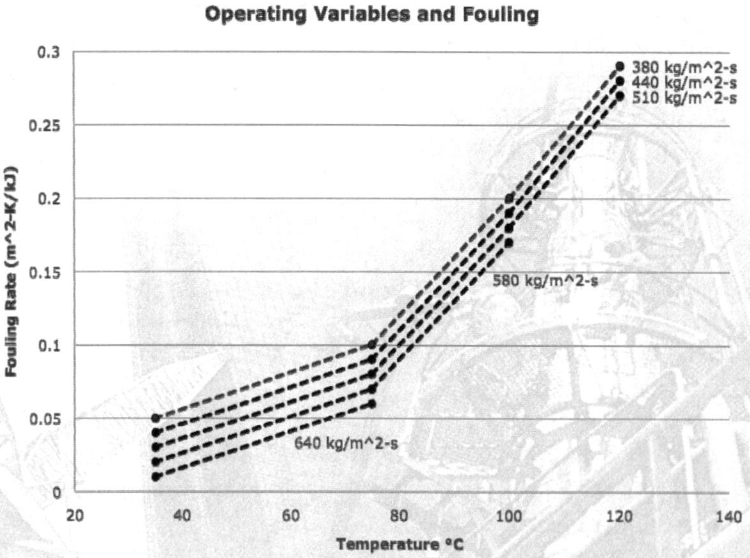

Cooling Water Fouling Costs

The costs summarized in the previous slide are based on detailed cost assessment made by Van Nostrand, Leach and Haluska for NATO countries.

They are comprised of energy losses, throughput, and maintenance, whose percentages are broken down.

Throughput is directly related to energy savings – increasing throughput decreases the energy cost per lb of product produced.

Cost Component	Percent of Annual Cost
Energy	34.8%
Throughput	64.1%
Maintenance & Cleaning	1.1%

Cooling Water Fouling Cost Basis

Type of fouling	Usual extent of problem
Chemical reaction (process side)	Major
Crystallization (cooling water)	Medium
Particulate (cooling water, gas side)	Minor/medium
Biofouling (cooling water)	Medium
Corrosion (process side)	Medium

Extent of Problem	Major	Medium	Minor/medium
Percent weight variable	X	Y	Z
Occurrence/Frequency	1	3	1
Extent of Problem	Major	Medium	Minor/medium

$X + 3Y + Z = 100$ & $X=3Y$, $Y=2Z$: Solving, $X=46.14\%$, $Y=15.39\%$, $Z=7.69\%$

Cooling Water Fouling Percentage

- Cooling water fouling occurs due to crystallization, particulate, and biofouling.
- Crystallization and biofouling occur in cooling water alone. Their extents are both medium. Thus, these two types of fouling contribute to 15.39 ×2 %=30.78% of total fouling costs.
- Particulate fouling is present in both gas side and cooling water. It is assumed that 60% of particulate fouling is due to cooling water. The extent of particulate fouling is minor/medium = 7.69%. Thus, particulate fouling in cooling water contributes to 7.69×0.6 %=4.614%.
- Adding these values together, the percent of total fouling that is caused by cooling water systems is 30.78+4.614=35.394% ≈35%.

Cooling Water Fouling Costs/Savings

Refinery	Refining Capacity (Barrels/Day)	Cooling Water Fouling Cost (millions of $)	Potential Annual CW Savings (millions of $)
ExxonMobil Refining & Supply Co Baytown	520,975	16.28	4.07
ExxonMobil Refining & Supply Co Baton Rouge	457,730	14.04	3.51
BP Products North America Inc Texas City	414,769	12.52	3.13
CITGO Petroleum Corp Lake Charles	390,845	11.67	2.92
BP Products North America Inc Whiting	368,550	10.88	2.72
ExxonMobil Refining & Supply Co Beaumont	313,495	8.93	2.23
Sunoco Inc (R&M) Philadelphia	304,850	8.63	2.16
Chevron USA Inc Pascagoula	300,300	8.46	2.12
Deer Park Refining Ltd Partnership Deerpark	300,118	8.46	2.11
WRB Refining LLC Wood River	278,460	7.69	1.92
Flint Hills Resources LP Corpus Christi	262,506	7.13	1.78
Premcor Refining Group Inc Port Arthur	261,170	7.08	1.77
Motiva Entreprises LLC Port Arthur	259,350	7.02	1.75
Flint Hills Resources LP Saint Paul	255,255	6.87	1.72
Chevron USA Inc El Segundo	253,890	6.82	1.71
Houston Refining LP Houston	246,246	6.55	1.64
BP West Coast Products LLC Los Angeles	241,150	6.37	1.59
Marathon Petroleum Co LLC Garyville	232,960	6.08	1.52
ConocoPhillips Company Belle Chasse	224,770	5.79	1.45
ConocoPhillips Company Sweeny	224,770	5.79	1.45

Chevron Cooling Water Study

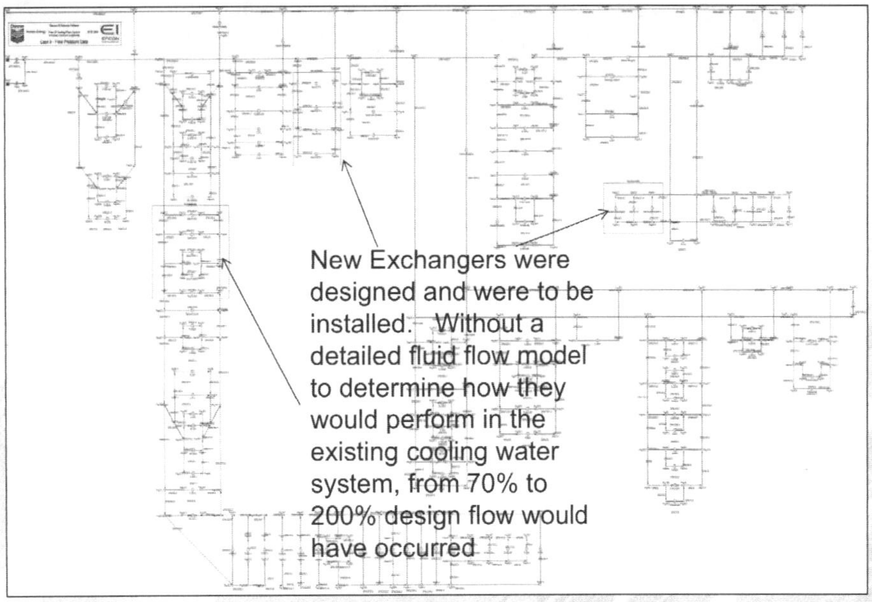

Fluid Flow Models Allow Tuning

- Install flow balancing valves/orifices to reduce flow to exchangers significantly exceeding their design flow and redistribute this flow to those identified significantly below their design flow. These flow balancing devices can be simulated and accurately sized using a fluid flow model.
- Reconfigure exchangers that are below their design flow from series to parallel flow where possible. These modifications can be simulated in detail using a fluid flow model.
- Change the layout of the cooling water distribution system to remove booster pumps that may be starving local exchangers and providing more pumping capacity at the beginning of the system and effectively make pipe sizes larger without replacing the existing pipe in operation through the use of cooling water flow simulation software.
- Evaluate splitting off exchangers from one cooling water tower system to another system when flow capacity is available in another cooling water tower system which can be analyzed in detail when fluid flow models are available for both systems.

Other Significant, Long Term Losses

- Extra heat exchanger area is required which actually doubles the capital costs, apart from making the heat transfer less effective (15).
- Oftentimes, to avoid extra area, redundant equipment is installed, such as two heat exchangers to perform the job that one should be able to achieve.
- Also, costly specialty materials and geometries are sometimes used (4).
- Extra energy is consumed due to the increase in pressure drop that results from fouling. This drastically increases heat transfer inefficiency and results in higher pumping costs (1).
- Fouling can also cause increased roughness in pipes, even obstructing the flow, or at the very least reducing it (15) .
- The quality of the water itself can be altered as well, making the current chemical treatment futile, and can lead to more corrosion of parts (14).

Other Significant, Long Term Losses

- This results in production losses thanks to extra plant shutdowns, maintenance, and cleaning.
- Additionally, heat exchanger parts must constantly be replaced due to condenser tube leakages, mechanical damage, and pump vibration.
- All of these factors lead to the impairment of the very integrity of the entire system and safety (15) (14).
- Additional cost would occur with the new Carbon Tax which is coming applicable in 2010. These costs have not included in the savings but can be significant.

Conclusions & Recommendations

- The various types of fouling often occur simultaneously and result in an increase in overall fouling, a greater heat loss, increased pressure drop, and a lessened efficiency (1) (3).
- By understanding each contributor to a heat exchanger's fouling and the effect of various parameters on the severity of fouling, industry can hope to begin to control this problem that has spiraled out of control since the invention of heat transfer equipment.
- The first step in this process is to identify where the problem is occurring in order to begin the remediation phase. A simple way to do this is to compare the design flows with the actual flows.
- Even if the very conditions of the process may cause fouling, one of the best ways to overcome or reduce it is to maintain design conditions (1).

Conclusions & Recommendations

- Such an analysis can be completed by performing a cooling water study, whose intent is exactly that.
- Once that is complete, it will be very evident as to where flows need to be changed in order to be in the optimum range for velocities that minimize fouling, as well as to control the surface temperature of the exchanger.
- Velocities should be kept as high as possible with an acceptable pressure drop and no other bad side effects such as erosion.
- Temperature should be monitored and controlled carefully with the flow as a temperature difference of just 5 °C can increase fouling by 100% (1).
- Indeed, fouling can be mitigated simply by maintaining an appropriate temperature and flow range during operation, two methods that are currently not commonly employed by industry (4).

In Summary...

- Cooling water heat exchanger fouling is a complex problem which involves design of the system and individual heat exchangers, operations, maintenance, and cooling water treatment programs.

- A cooling water fluid flow model compares "what it is" (current operations) with "what it should be" (design).

- This allows plant engineers and operators to pinpoint root cause problems of low cooling water flow and develop solutions to achieve design flow and thereby reduce the significant cost of cooling water system fouling.

References

1. **Bott, T.R.** Heat Exchanger Fouling and Its Mitigation. [book auth.] E.A. Foumeny and P.J. Heggs. *Heat Exchange Engineering Volume 2 Compact Heat Exchangers: Techniques of Size Reduction.* Chichester : Ellis Horwood Limited, 1991.
2. **Smith, Malcolm, Jenkins, Andrew and Grant, Colin.** Comparison of fouling data from alternative cooling water sources. [book auth.] Naim H. Afgan, Zeljko Bogdan and Neven Duic. *Sustainable Development of Energy, Water and Environment Systems.* Lisse : Swets & Zeitlinger Publishers, 2004.
3. **Nunn, Robert G.** *Water Treatment Essentials for Boiler Plant Operation.* New York : McGraw-Hill, 1997. 0-07-048219-5.
4. **Garret-Price, B.A., et al.** *Fouling of Heat Exchangers Characteristics, Costs, Prevention, Control, and Removal.* Park Ridge : Noyes Publishers, 1985. 0-8155-1016-0.
5. **Knudsen, James G.** Fouling in heat exchangers. *Heat Exchanger Design Handbook.* Washington : Hemisphere Publishing Corporation, 1983.
6. **Hays, G.F.** Fouling Allowances are Self-Fulfilling Prophecies. [book auth.] R.K. Shah. *Heat Transfer Equipment Fundamentals, Design, Applications, and Operating Problems.* New York : The American Society of Mechanical Engineers, 1989.
7. **Watkinson, Paul.** Cooling Water Fouling - Summary of Discussion. [book auth.] L.F. Melo, T.R. Bott and C.A. Bernardo. *Fouling Science and Technology.* Dordrecht : Kluwer Academic Publishers, 1988.
8. **Shah, Ramesh K. and Sekulic, Dusan P.** *Fundamentals of Heat Exchanger Design.* Hoboken : John Wiley & Sons, Inc., 2003. 0-471-32171-0.
9. **Webb, Ralph L. and Kim, Nae-Hyun.** *Principles of Enhanced Heat Transfer.* New York : Taylor & Francis Group, 2005. 1-59169-014-5.
10. **Pritchard, A.M.** Deposition of Hardness Salts. [book auth.] L.F. Melo, T.R. Bott and C.A. Bernardo. *Fouling Science and Technology.* Dordrecht : Kluwer Academic Publishers, 1988.
11. **Foumeny, E.A. and Heggs, P.J.** Advanced Heat Exchangers - Will Fouling Be A Problem? [book auth.] E.A. Foumeny and P.J. Heggs. *Heat Exchange Engineering Volume 1 Design of Heat Exchangers .* Chichester : Ellis Horwood Limited, 1991.
12. **Crittenden, B.D.** Chemical Reaction Fouling of Heat Exchangers. [book auth.] L.F. Melo, T.R. Bott and C.A. Bernardo. *Fouling Science and Technology.* Dordrecht : Kluwer Academic Publishers, 1988.

References

13. **Chan, S.H. and Ghassemi, K.F.** Analytical Modeling of Calcium Carbonate Deposition for Laminar Falling Films and Turbulent Flow in Annuli: Part I - Formulation and Single-Species Model. [book auth.] R.K. Shah. *Heat Transfer Equipment Fundamentals, Design, Applications, and Operating Problems.* New York : The American Society of Mechanical Engineers, 1989.

14. **Nair, K.V. K., Satpathy, K. K. and Venugopalan, V. P.** Biofouling Control: Current Methods and New Approaches with Emphasis on Power Plant Cooling Water Systems. [book auth.] Rachakonda Nagabhushanam and Mary-Frances Thompson. *Fouling Organisms of the Indian Oceanq.* Rotterdam : A.A. Balkema, 1997.

15. **Melo, L. F. and Pinheiro, M. M.** Biofouling in Heat Exchangers. [book auth.] L. F. Melo, et al. *Biofilms - Science and Technology.* Dodrecht : Kluwer Academic Publishers, 1992, pp. 499-509.

16. **Pinheiro, M.M.V.P.S., et al.** Effect of Design and Operating Variables on Biofouling. [book auth.] L.F. Melo, T.R. Bott and C.A. Bernardo. *Fouling Science and Technology.* Dordrecht : Kluwer Academic Publishers, 1988.

17. **Bott, T. R.** Industrial Monitoring - Cooling Water Systems. [book auth.] L. F. Melo, et al. *Biofilms - Science and Technology.* Dordrecht : Kluwer Academic Publishers, 1992.

18. **Roques, Henri.** *Chemical Water Treatment Principles and Practice.* Paris : VCH Publishers, Inc., 1996. 1-56081-518-3.

19. **Seneviratne, Mohan.** *A Practical Approach to Water Conservation for Commercial and Industrial Facilities.* Jordan Hill : Elsevier Ltd., 2007. 978-1-85-617489-3.

20. **Energy Information Administration.** Annual U.S. Percent Utilization of Refinery Operable Capacity . *Energy Information Administration - EIA - Official Energy Statistics from the U.S. Government.* [Online] June 29, 2009. [Cited: August 4, 2009.] http://www.eia.doe.gov/.

21. **BP.** Conversion Factors. *BP Global BP.* [Online] 2009. [Cited: August 4, 2009.] http://www.bp.com/.

22. **Energy Information Administration.** Top U.S. Refineries - Energy Information Administration. Energy Rankings. *Energy Information Administration - EIA - Official Energy Statistics from the U.S. Government.* [Online] July 2008. [Cited: August 5, 2009.] http://www.eia.doe.gov/.

23. **Chenoweth, James M.** General Design of Heat Exchangers for Fouling Conditions. [book auth.] L.F. Melo, T.R. Bott and C.A. Bernardo. *Fouling Science and Technology.* Dordrecht : Kluwer Academic Publishers, 1988.

24. **Donlan, Rodney M.** Biofilm Control in Industrial Water Systems: Approaching an Old Problem in New Ways. [book auth.] L.V. Evans. *Biofilms: Recent Advances in their Study and Control.* Amsterdam : Harwood Academic Publishers imprint, 2000.

References

25. **Thompson, Ken, et al.** *Characterizing and Managing Salinity Loadings in Reclaimed Water Systems.* Denver : Awwa Research Foundation, 2006.

26. **Rabas, T.J., et al.** Comparison of Power-Plant Condenser Cooling-Water Fouling Rates for Spirally-Indented and Plain Tubes. [book auth.] T.J. Rabas and J.M. Chenoweth. *Fouling and Enhancement Interactions.* New York : The American Society of Mechanical Engineers, 1991.

Hydrogen Use in Steam- and Air-Assisted Flares

Alexis McKittrick, Uyi Iyoha, Euan Evenson, Lee Rosen, Lynne Dant

Praxair, Inc.

AIChE Process Technology Conference, Galveston, TX

October 2, 2009

www.praxair.com

Hydrogen as a Fuel for Flaring

- Current EPA flare compliance options
  - Non-assisted Flares:
    - 200 Btu/scf minimum OR
    - 8 volume % hydrogen minimum
  - Assisted flares
    - 300 Btu/scf minimum
- Natural gas added to "sweeten" low-Btu streams and comply with regulations
- Much less hydrogen than natural gas is needed to comply with EPA requirements for low-Btu flare streams

- Praxair pursuing modification of 40 CFR 60.18 and 40 CFR 63.11 to extend the current specification to include steam- and air-assisted flares
- Potential benefits:
  - Lower operating cost - 50% to 67% reduction
  - Lower environmental emissions – CO_2, CO, NO_x
  - Lower net natural gas consumption

2

Low-Btu Flaring for Assisted Flares

PRAXAIR

Currently, many flare operators add natural gas to the waste stream to comply with 300 Btu/scf regulation

Using Praxair's proposed regulation, flare operators could add Hydrogen to 11.5 vol% to comply

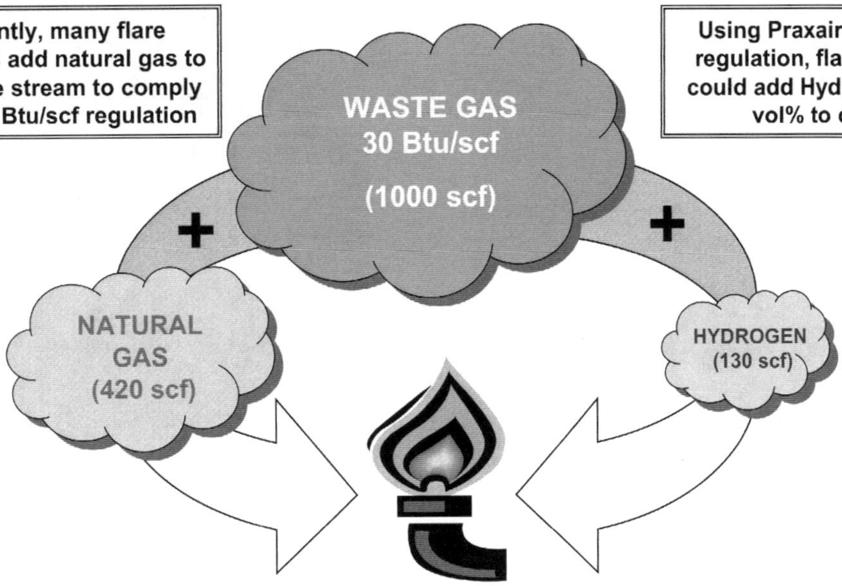

3

Benefits Example

PRAXAIR

Assisted Flare – 100k scfh stream @ 30 Btu/scf

	Natural Gas 300 Btu/scf	Hydrogen 11.5 vol.%	Reduction (%)
Fuel gas required for compliance (scfh)	42,000	13,000	69
Annual fuel costs (24/7 operation, 340 d/y) *$4/MMBtu Natural Gas*	$1,400,000	$425,000	70
NOx emissions (ton/yr)	11.8	1.9	84
CO emissions (ton/yr)	60.2	9.9	84
CO_2 emissions (ton/yr)	19,363	2,605	87

~$1MM/year in potential annual savings

4

Testing Hydrogen Addition to Assisted Flares

PRAXAIR

- Flare testing facility built in Tonawanda, NY – 1H 2008
  - Variety of fuels
  - Assist / non-assist operation
- <u>Goal:</u> Add hydrogen compliance option for assisted flares to EPA regulations
- Critical parameters:
  - >98% destruction efficiency of hydrocarbons
  - Flame stability
  - No visible emissions / smoke

5

Praxair Steam/Air-Assisted Flare Test Facility

PRAXAIR

6

Test Variables for Assisted Flares

PRAXAIR

Variable	Range / Value Tested
H2 volume %	5 - 25%
Hydrocarbon heating value	30, 100, & 200 Btu/scf
Velocity	20 - 150 ft/s
Steam Assist	0.4 lb steam / lb ethylene
Air Assist	20% of stoichiometric air requirement

7

Flame Stability Images

PRAXAIR

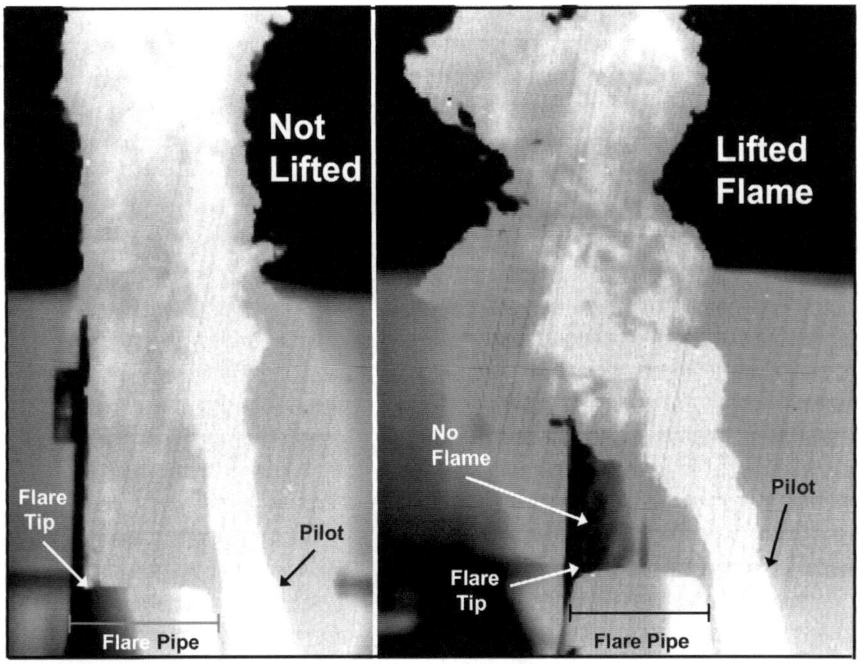

8

Flame Stability Results – Air- and Steam-Assisted Flares

PRAXAIR

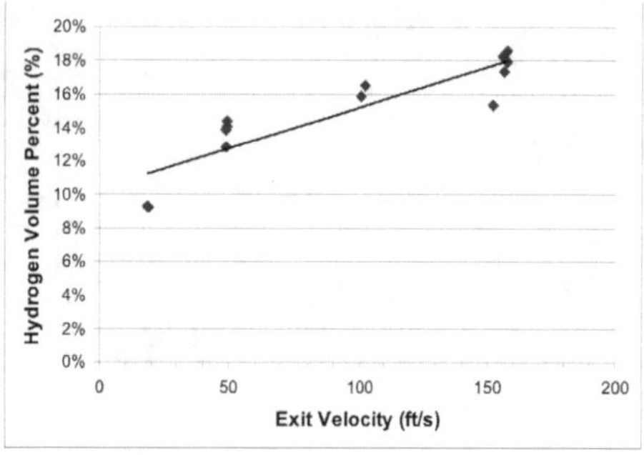

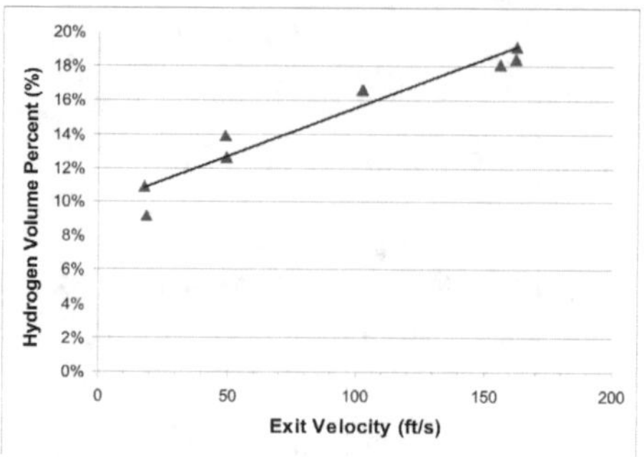

Fitting linear equations to the flame stability data to determine the amount of hydrogen required to ensure a stable flame at a given exit velocity

9

Destruction Efficiency Results Using Hydrogen

PRAXAIR

Heating Value of Waste Gas (Btu/scf)	*Assist Used*	*Runs n*	*Destruction Efficiency (%) Average*	*Destruction Efficiency (%) 95% Confidence*	*Exit Velocity (ft/s)*	*Flame Stability Ratio*
30	Steam	3	99.77	**99.61**	151	1.03
100	Steam	4	99.94	**99.92**	150	1.04
200	Steam	3	99.99	**99.98**	151	1.03
30	Air	4	99.26	**99.02**	150	1.02
100	Air	4	99.90	**99.88**	150	1.02
200	Air	3	99.92	**99.90**	150	1.02

Destruction efficiency > 99% measured for all tests

10

Commercial Installation

- Flare converted to H_2-supplemented fuel **June 2009**
- Nominal flare gas flowrate: **120,000 scfh**
- Supplemental H_2 concentration: **8 vol%**
- Nominal H_2 flowrate: **10,000 scfh**
- H_2 supply mode: **Pipeline**

11

Commercial Installation (Cont.)

- **Capital Investment**
  - H_2 flow control skid
  - Piping to and from skid
  - H_2 tank (for liquid H_2)
  - Concrete pad
  - Vaporization equipment

- 3 to 4 months authorization to on stream

H_2 flow control skid

12

Customer Benefits

PRAXAIR

Non-assisted Flare – 120k scfh stream @ 30 Btu/scf

	Natural Gas 200 Btu/scf	Hydrogen 8 vol.%	Reduction (%)
Fuel gas required for compliance (scfh)	30,000	10,500	65
Annual fuel costs (24/7 operation, 340 d/y) *$5/MMBtu Natural Gas*	$1,250,000	$430,000	66
Estimated NOx emissions (ton/yr)	7.9	1.7	79
Estimated CO emissions (ton/yr)	67.3	14.6	78
Estimated CO_2 emissions (ton/yr)	13,831	2,105	85

~$820,000/year annual savings @ $5/MMBtu NG

13

NOx & CO Calculations: Based on Standard Emissions Index from TCEQ's Air Permit Technical Guidance for Chemical Sources: Flares and Vapor Oxidizers, October 2000, RG 109 (Draft) pg 18-19

Conclusions

PRAXAIR

- Hydrogen-supplemented flare application commercialized for non-assisted flares
  - Reduced flare operating expense by >~65%
  - Resulted in >98% combustion efficiency
  - Reduced natural gas consumption
  - Reduced emissions
- Completed pilot scale testing of hydrogen use in assisted flares:
  - Demonstrated >98% combustion efficiency
  - Demonstrated stable combustion
  - No visible emissions
- Requesting that 40CFR 60.18 & 63.11 be amended to include hydrogen as a compliance option for steam- and air-assisted flares based on Praxair testing results

14

15

Steam-Assisted Flares – Emissions Reduction with Hydrogen

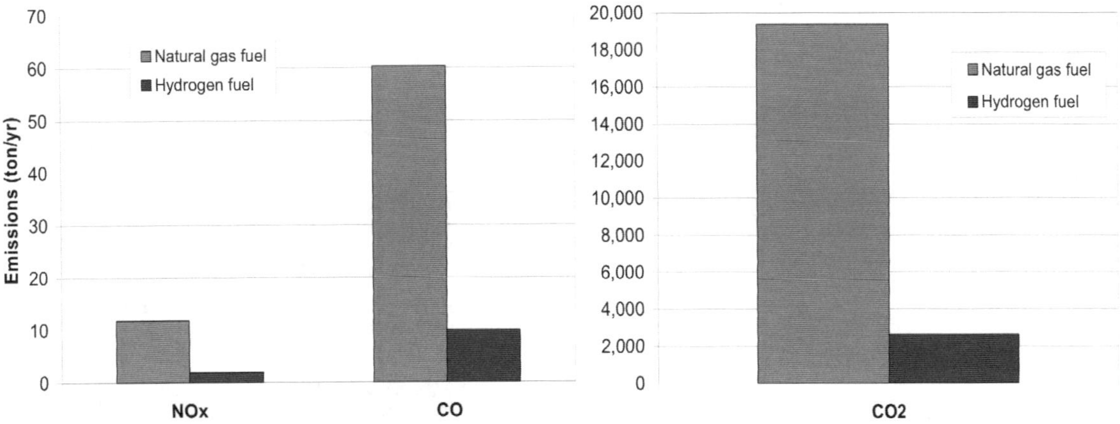

Stream assumptions: waste gas volume = 100,000 scfh; waste gas heating value = 30 BTU / scf

NOx & CO Calculations: Based on Standard Emissions Index from TCEQ's Air Permit Technical Guidance for Chemical Sources: Flares and Vapor Oxidizers, October 2000, RG 109 (Draft) pg 18-19

Thermal NOx – 0.068 lb/MMBTU for steam assisted

CO - 0.3465 lb/MMBTU for steam assisted

CO2 Calculations: Based on 9.5lbs of CO2 for 1lb H2 produced from SMR; 113lbs CO2 for 1000scf of Natural Gas

16

Optimizing Fired Heaters Operation Saves Money

- Ashutosh Garg, Furnace Improvements

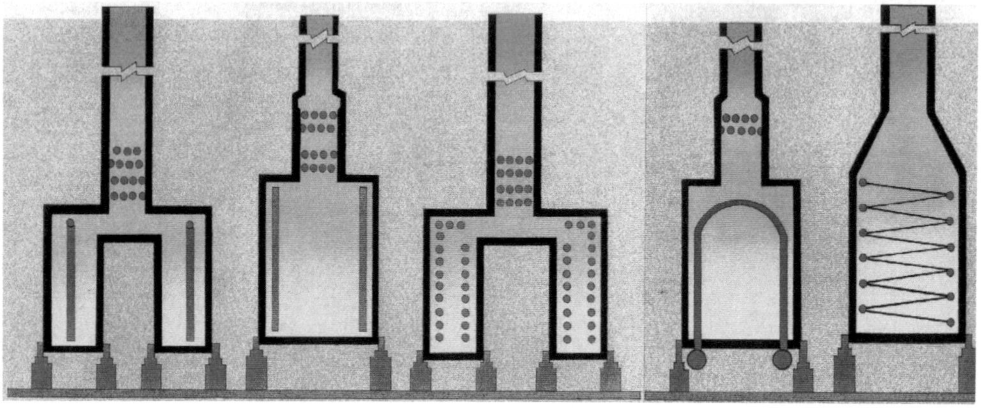

Energy Consumption

- Refinery
  - 0.44 MMBtu/ BBL of oil
- Ethylene
  - 22 MMBtu/ ton
- Ammonia
  - 28.5 MMBtu/ton

Energy Consumption

- A Typical refinery process 100,000 BPD
- Consumption of 0.40 MMBTU/BBL
- $6 per MMBTU
- Energy bill of $ 87.6 million per year.

Potential Saving

- 1% Efficiency improve → $876,000/Year

Fired Heaters

- Essential component in Petrochemical and Chemical Plants
- Each refinery has 20-50 fired heaters

Optimum Heater Operation

- Fired Heater:
  - Heating coil
  - Enclosure
  - Combustion Equipment
- Instrumentation controls:
  - Fuel Firing Rate
  - Flow Through coils

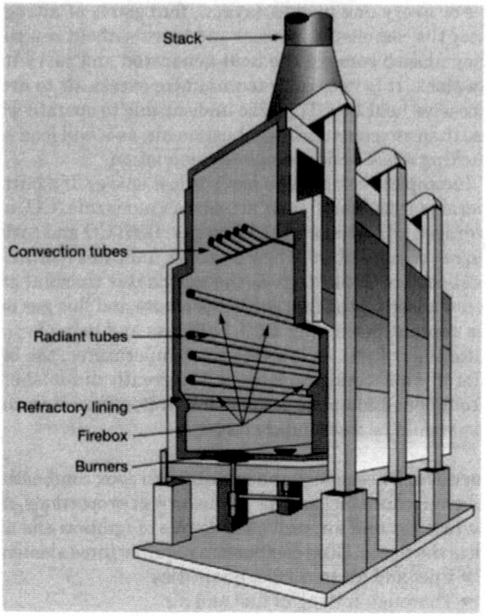

Combustion

- Combustion Requires
  - Air ($0.21O_2 + 0.79\ N_2$)
  - Fuel
  - Ignition source
- Incomplete Combustion
  - Energy Loss
  - CO and H_2 are released
- Burners
  - Mix correct quantities of fuel & air
  - Provide a source of ignition
  - Stabilize the flame

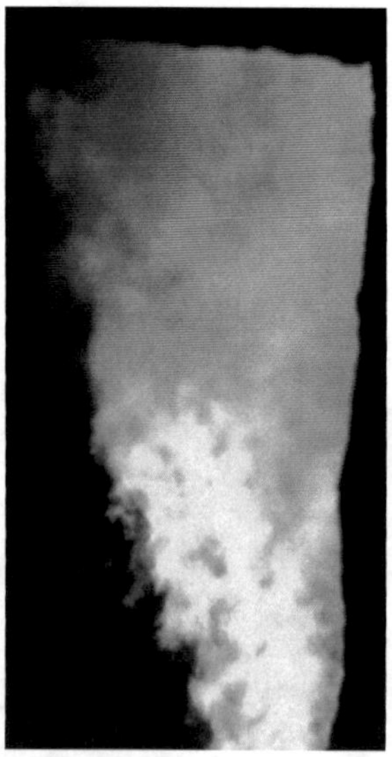

Draft

- Negative pressure inside the heater.
  - Combustion air is drawn inside the heater
  - Hot Flue gas flows out of the stack
- Types of Draft
  - Natural Draft
  - Forced Draft (FD)
  - Induced Draft(ID)
  - Balance Draft

Draft

- Dampers helps control the draft
- Arch is the highest pressure point in the heater
- Excess or shortage of draft is not acceptable

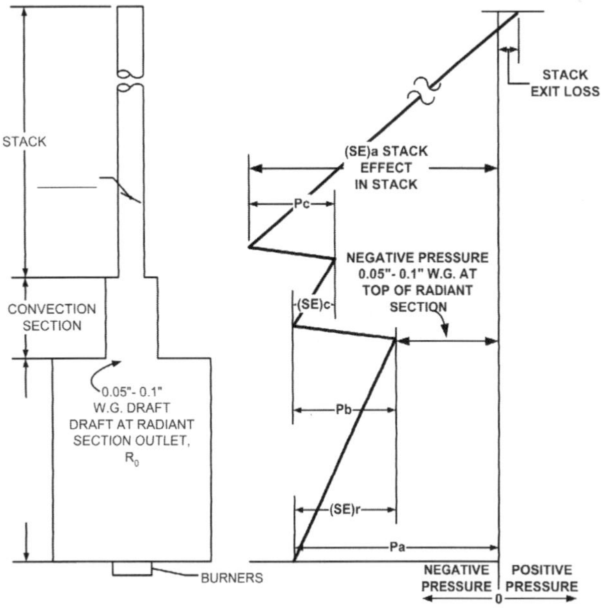

Burner Operation

- Correct combustion in firebox include:
  - The firebox is clear
  - No smoky appearance
  - Burners flames are steady and well-formed
- Check burners regularly
- Adjusting burner registers to control air intake

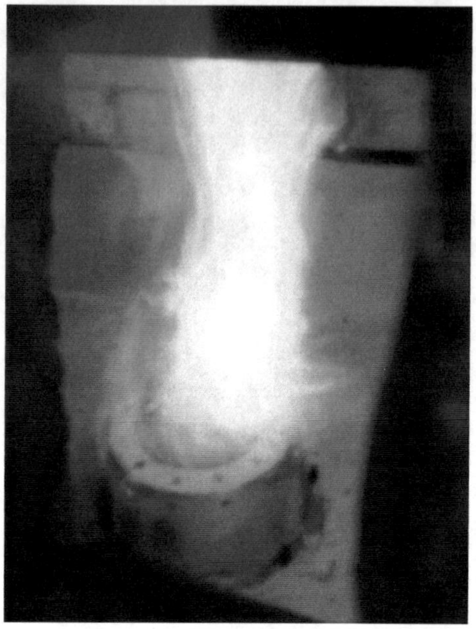

Air Leakages

- Air entering in to furnace from a number of places:
  - Peepholes
  - Header box doors
  - Tube guide opening
  - Unused burners

Heater optimization

- Target set 2-3% O2 in the flue gas as a target.
- Operating conditions fluctuates :
- Manual control /adjustments
- Operators
  - Number going down
  - Experience going down
  - Need training

Optimisation Case Study -1

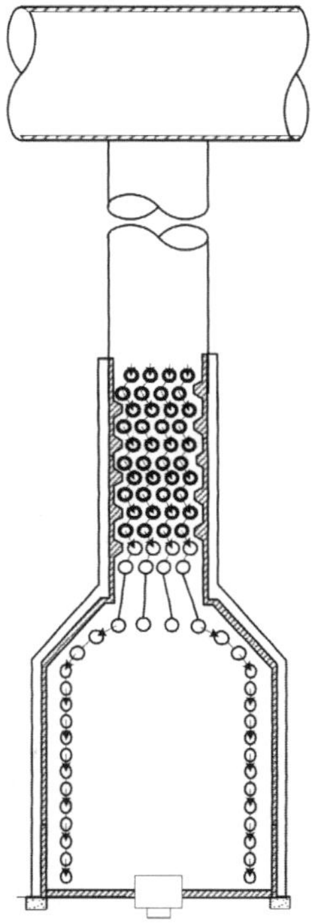

- FIS performed a tuning job for a refinery.
- Heater:
  - Depentanizer reboiler heater
  - Horizontal tube box
  - Absorbed heat duty - 87 MMBtu/hr
  - 15 up fired burners
  - Heater connected to a large common stack
  - Two off take ducts provided with manual dampers

Optimisation Approach

- Check draft
  - Adjust using off-take dampers
- Check excess O2
  - Adjust burner register
- Check burners
  - Light up all burners

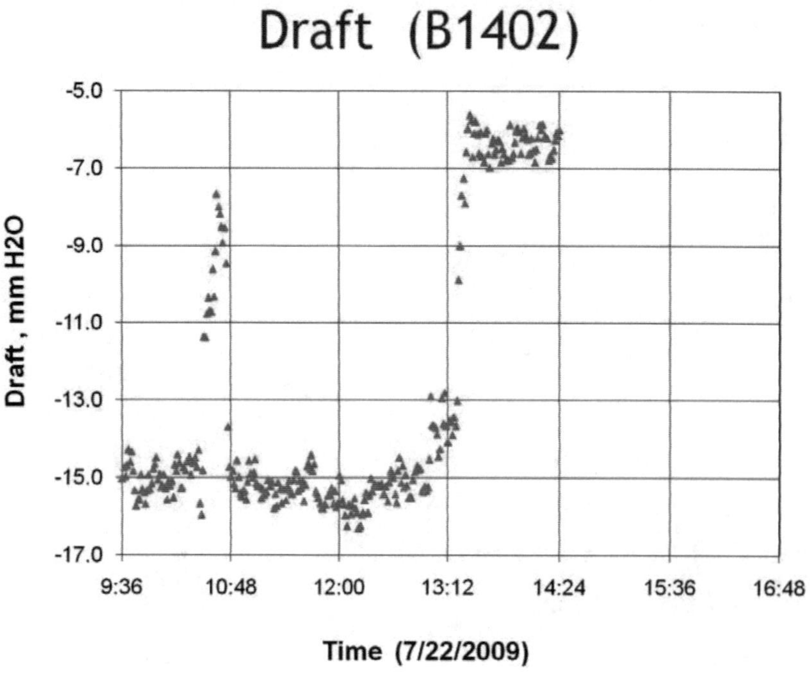

Excess Oxygen (B1402)

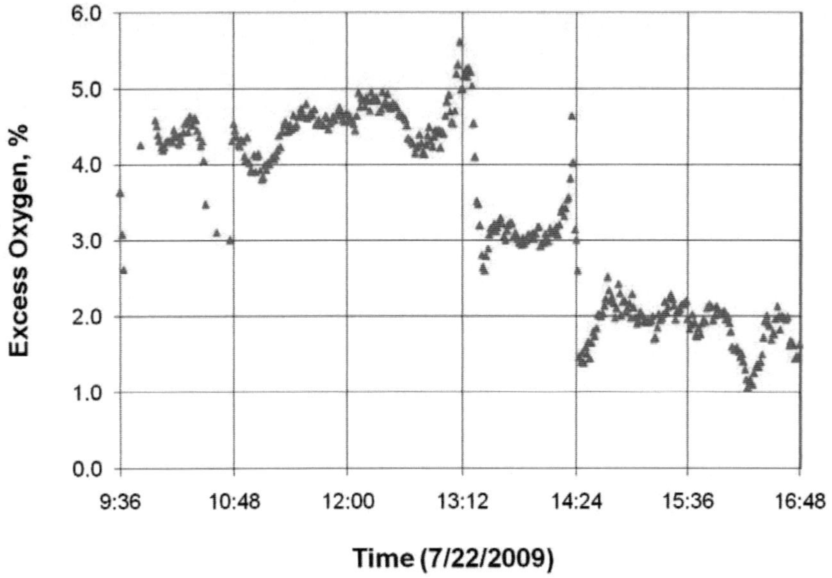

Stack Temperature (B1402)

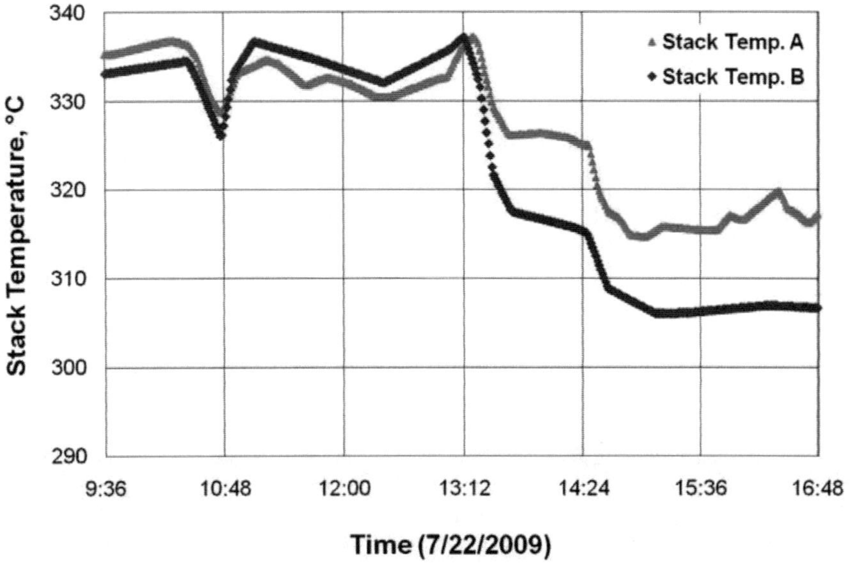

Thermal Efficiency (B1402)

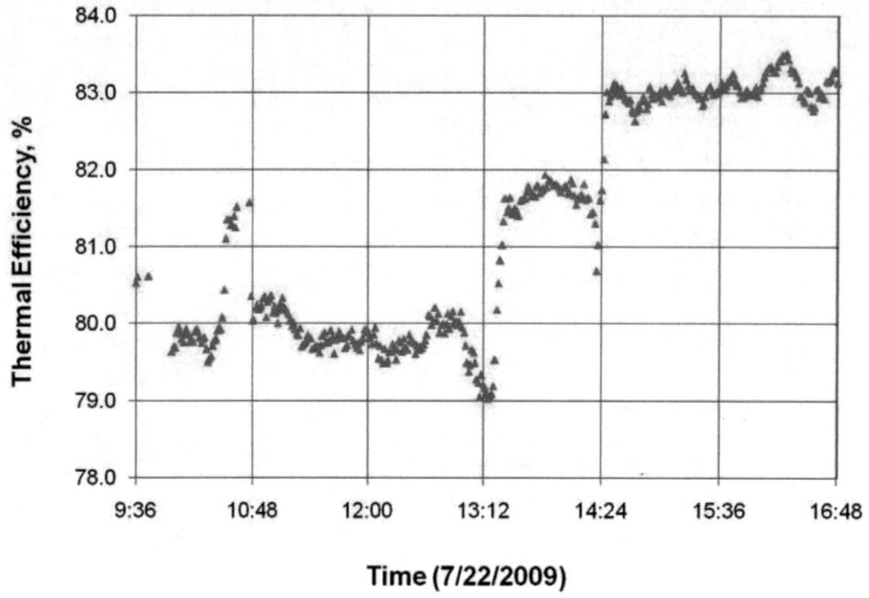

Total Heat Fired (B1402)

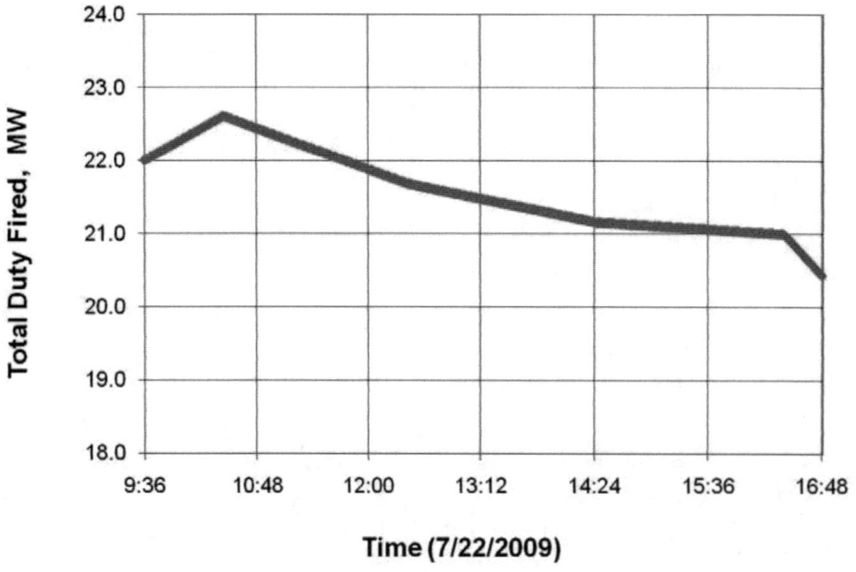

Crude Heater Tuning Case Study -2

- Natural Draft Crude Heater
- Horizontal tube
- Up-fired burners
- 11 burners

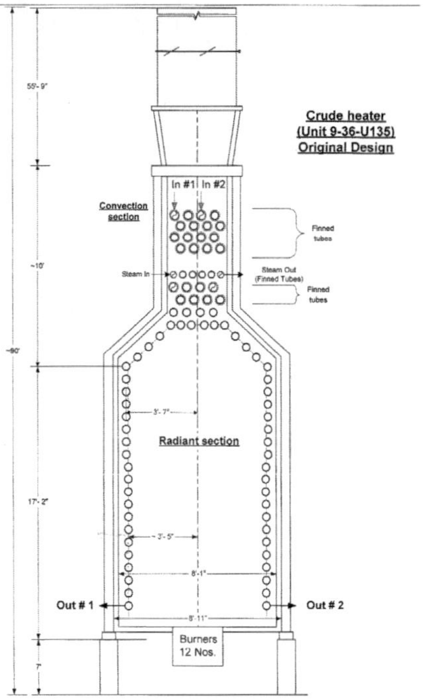

Fuel gas Flow and Pressure Before & After Tuning

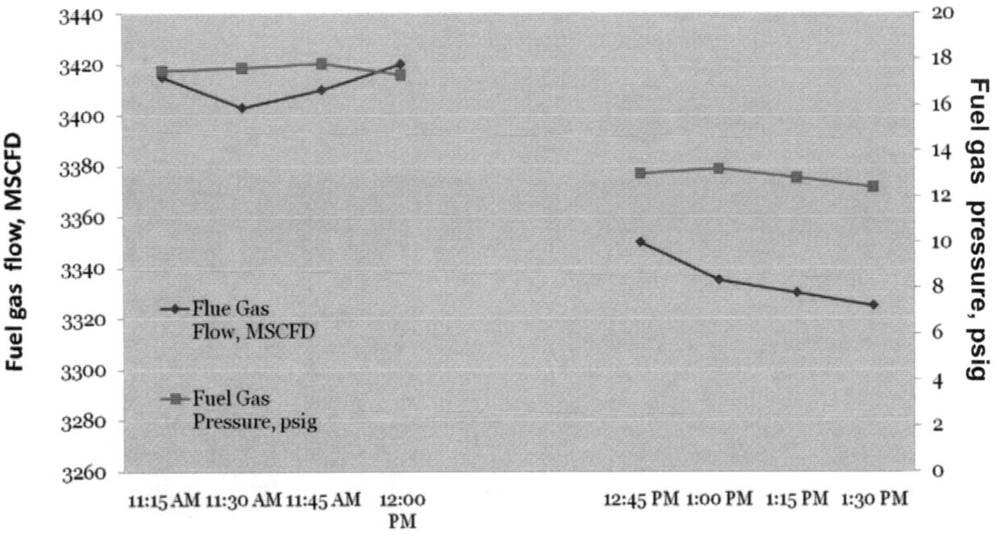

Tube Skin Temp Before & After

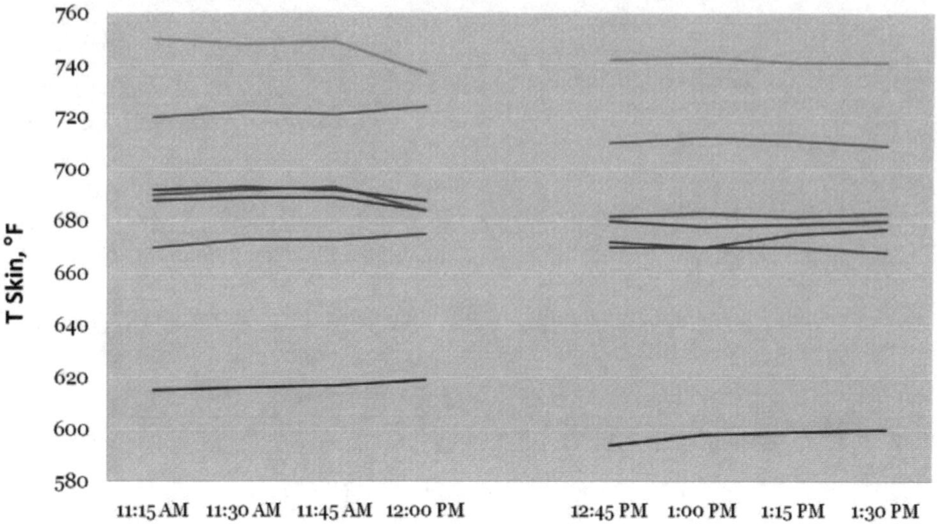

Recommendations

- Software
  - Heater Performance Index (HPI)
- Hardware
  - Draft control system
  - Burner control system

Heater Performance Index

- Analyzes the performance of Fired Heater 24/7
- Built into DCS
- Generates guidelines
- Built in intelligence
- Customized modeling

- Thermal Efficiency
- Fuel gas rate,
- Draft,
- Excess O2,
- Tube skin temperatures,
- Feed flow rate,
- Pressure drop,
- Coking rate,
- Air preheater performance

Heater Performance Index

Description	Flow (BPM)	Inlet Pressure (PSIG)	Crossover Temp (°F)	Outlet Temp (°F)	TMT (°F)
Pass #1	709.5	278.92	585.1	686.78	960.7 / 969.4
Pass #2	676.52	295.97	589.75	678.62	965.1 / 994.5
Pass #3	670.8	275.85	592.02	680.15	895.8 / 927.1
Pass #4	729.08	285.95	597.14	678.49	907 / 1034.4
Pass #5	729.1	281.66	594.54	683.37	930.6 / 1041.67
Pass #6	695.13	283.84	590.9	670.14	887.6 / 845.62
Pass #7	737.39	298.2	594.39	682.95	959.9 / 863.41
Pass #8	653.35	275.73	583.27	679.04	964 / 1048.68

No. of burners in operation	24
Heater Performance Index	0.56
Thermal Efficiency	90.25
Fuel Saving Potential	3.08

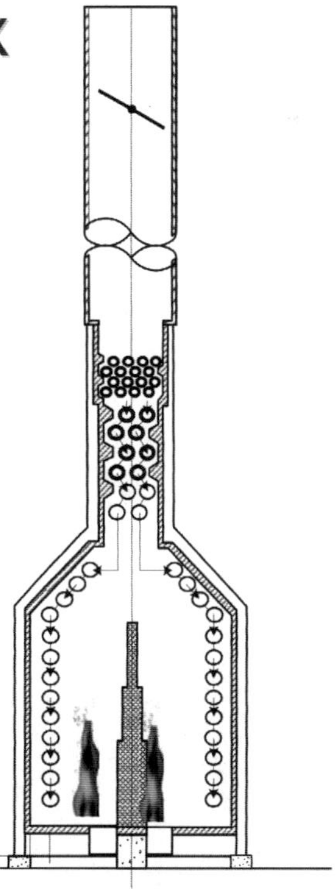

Performance Monitors

Individual pass flow above minimum acceptable
Heat pickup in convection section is high – Afterburning Suspected
Radiant coil metal temperature is OK
Burners are OK
Oxygen in firebox OK
Draft in the firebox is OK
Number of burners in operation is OK
Air Flow reported appears erroneous. Check the meter calibration.

Reliable Dampers

Opposed blades vs. Parallel Blades

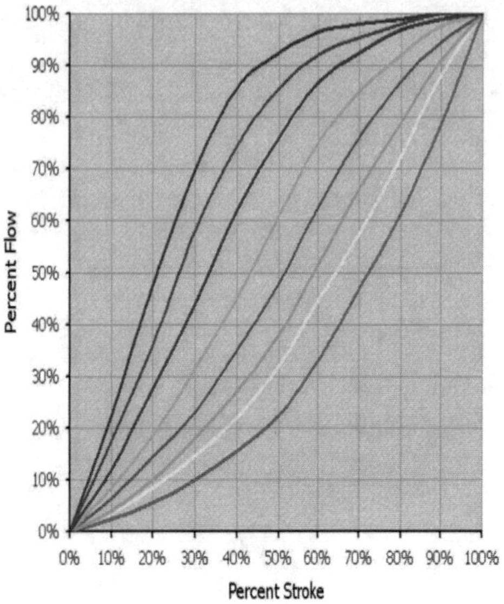

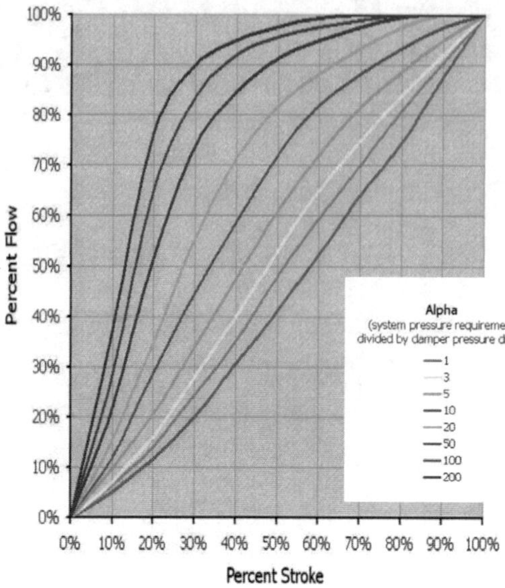

Oxygen & Draft Control System

(Natural Draft Heater)

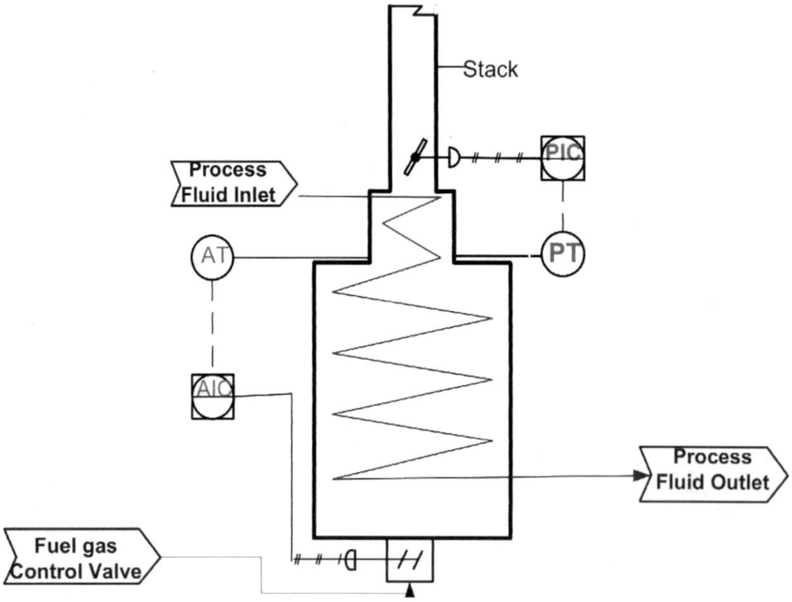

Thank you very much

- Questions and comments are welcome

MUSTANG

Analysis of Greenhouse Gas (GHG) Regulation Impacts to the US Refining Industry

Ian M. Glasgow, Stan Polcar, R.E. "Ed" Palmer

Regional AIChE Technology Conference
October 1-2, 2009

MUSTANG

Agenda

- **Background**
- **Enacted and Proposed Legislation**
- **Life Cycle Assessment (LCA)**
- **Refinery Case Studies**
  - **Low Carbon Fuel Standard**
  - **Cap and Trade**
- **Observations and Future Work**

Background

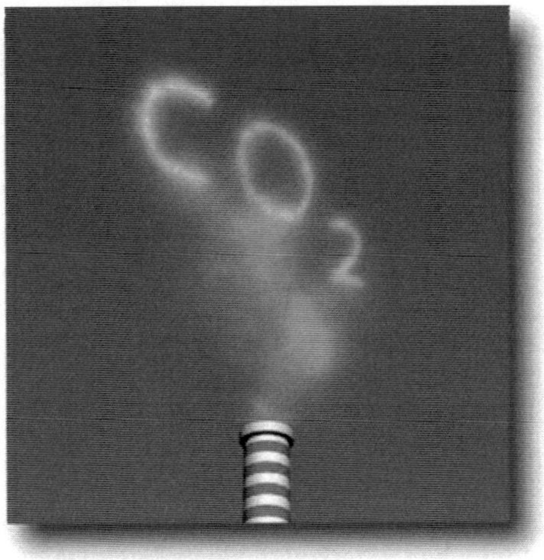

- GHG Emissions
  - CO_2
  - N_2O
  - CH_4
  - SF_6
  - HFC
  - PFC
- Remain concentrated over long period of time
- Not concentrated in area of emission
- Global not regional solution

Background

U.S. Greenhouse Gas Emissions, 2007

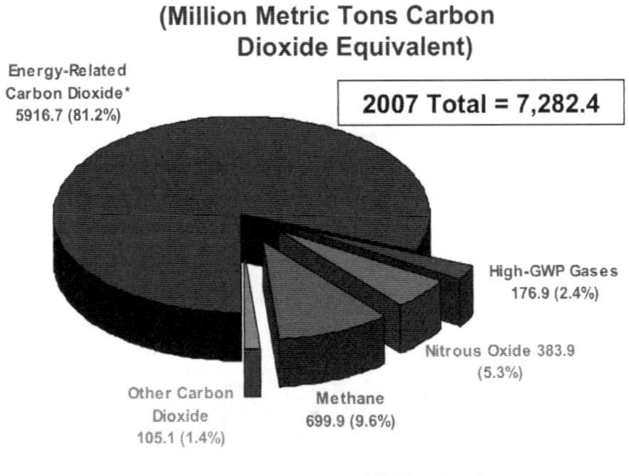

*Adjusted.

From EIA webpage; www.eia.doe.gov

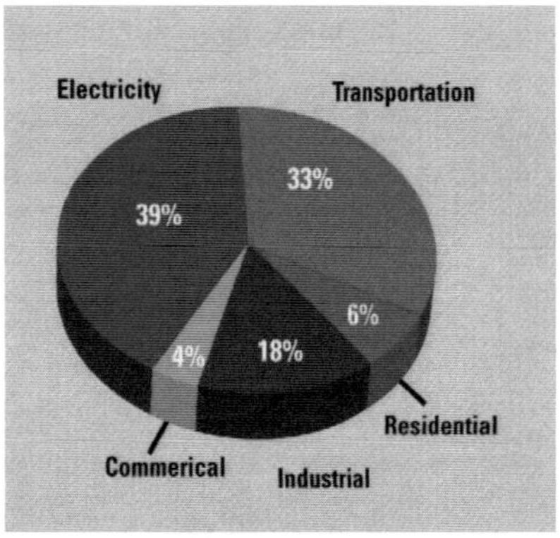

From www.pewclimate.org

Enacted Legislation

MUSTANG

- Energy Independence Act of 2007
  - Renewable Fuel Standard (RFS-2)

TOTAL BioFuel
36 Billion Gallon per year

Renewable Fuel	Advanced BioFuel	Cellulosic BioFuel
15 Billion Gallon per year	5 Billion Gallon per year	16 Billion Gallon per year
20% GHG Reduction	50% GHG Reduction	60% GHG Reduction
NET: 36 Bgal/yr	NET: 21 Bgal/yr	NET: 16 Bgal/yr

  - Increased CAFÉ standards

Enacted Legislation

MUSTANG

- Key Regulatory Announcements
  - Reporting GHG emissions from covered facilities
    - Reporting to being in 2010
  - 2007 Massachusetts v. EPA Supreme Court Decision
    - EPA has stated endangerment exists
    - EPA yet to determine how CAA will be used
  - Preliminary LCA Analysis of biofuels in RFS-2
    - Corn Ethanol does not meet renewable fuel definition with Indirect Land Change
  - CAFÉ standards increased
    - To include GHG in regulation
    - 30% reduction to 35.5 mpg light duty fleet
      - 39 mpg for cars; 30 mpg for trucks

Enacted Legislation

- California Low Carbon Fuel Standard (LCFS)
  - Reduction in Transportation fuel life cycle emissions
    - "Cradle to Grave" or "Well to Wheels"
  - 10% reduction of GHG from 2010 to 2020
    - 2010 is reporting only
    - 2015 – 2.5% reduction
    - 2020 – 10% reduction
  - Approved by CARB in April 2009
  - LCA emissions provided by CARB
    - Indirect Land Change included
  - Market driven with tradable credits
  - Oregon passed similar legislation July 2009
  - FL, MN, WA, and RGGI considering similar legislation

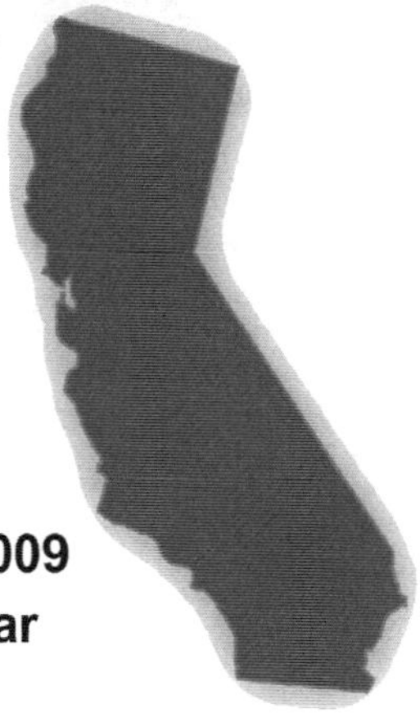

Proposed Legislation

- Waxman-Markey American Clean Energy and Security Act
  - Multi-phase program similar to California AB 32
    - Cap-and-Trade

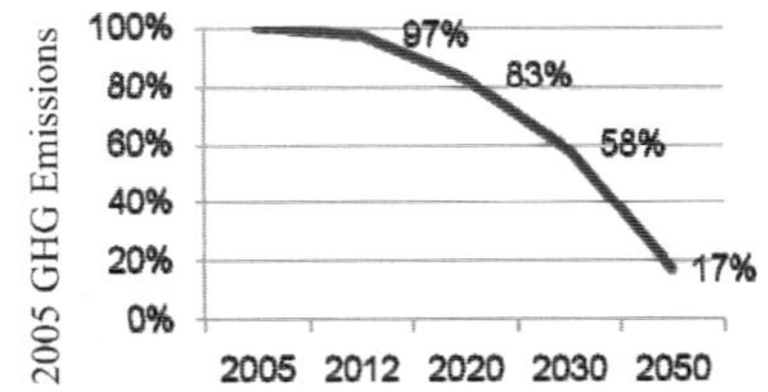

    - NO Low Carbon Fuel Standard
    - Vehicle emission standards for all transportation sectors
    - 85% Allowances for free to start
      - 2% to refining until 2026 (compared to ~ 35% of current total)
      - 2027 nearly all allowances are by auction
    - Emissions mandates of highest GHG and CH_4 emitters
      - To be determined by EPA at later date
    - Offsets receive 0.8 CO_2 equiv after first 5 years
      - Project to be defined by Advisory Board
      - Example: Heavy Industrial Power Reduction Projects

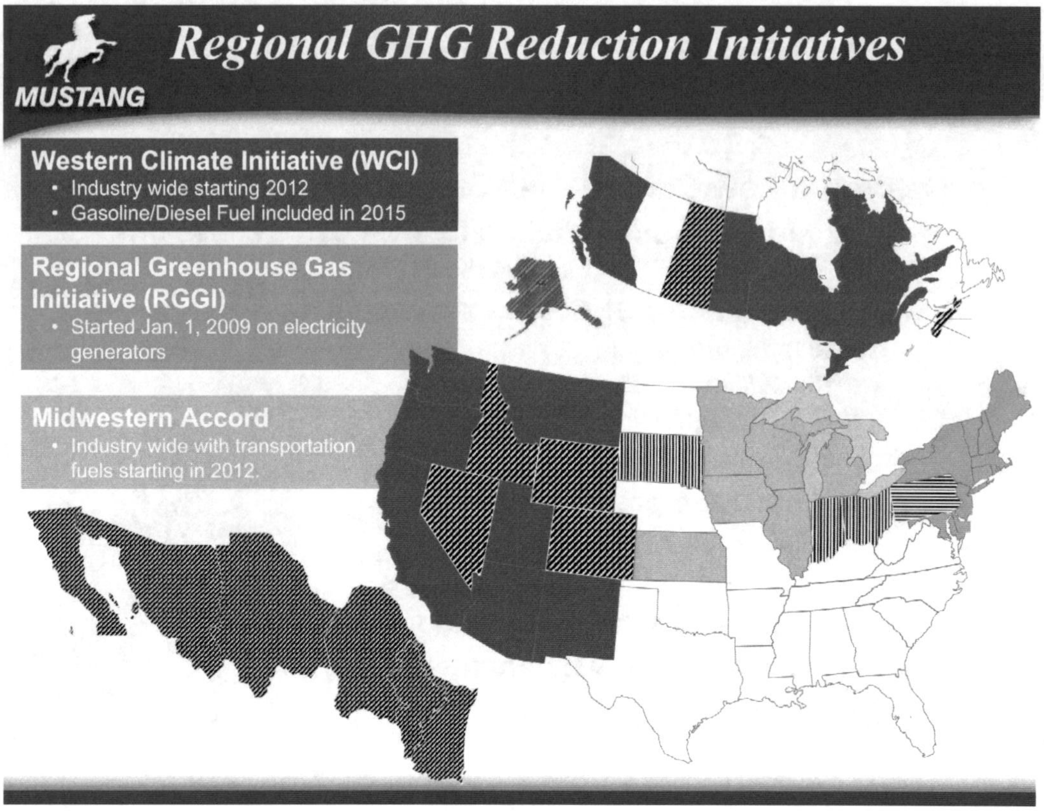
MUSTANG
Regional GHG Reduction Initiatives
Western Climate Initiative (WCI)
• Industry wide starting 2012
• Gasoline/Diesel Fuel included in 2015
Regional Greenhouse Gas Initiative (RGGI)
• Started Jan. 1, 2009 on electricity generators
Midwestern Accord
• Industry wide with transportation fuels starting in 2012.

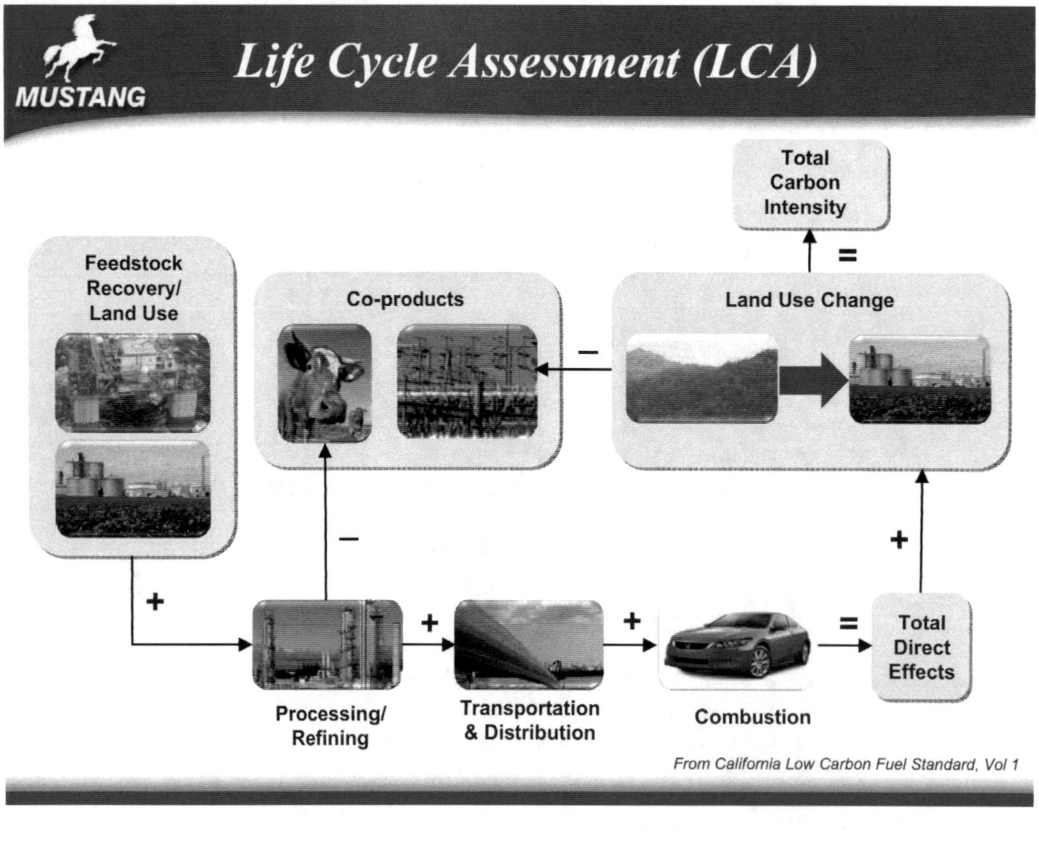
MUSTANG
Life Cycle Assessment (LCA)
Total Carbon Intensity
=
Feedstock Recovery/ Land Use
Co-products
Land Use Change
–
–
+
+
Processing/ Refining
+
Transportation & Distribution
+
Combustion
=
Total Direct Effects
From California Low Carbon Fuel Standard, Vol 1

Life Cycle Assessment (LCA)

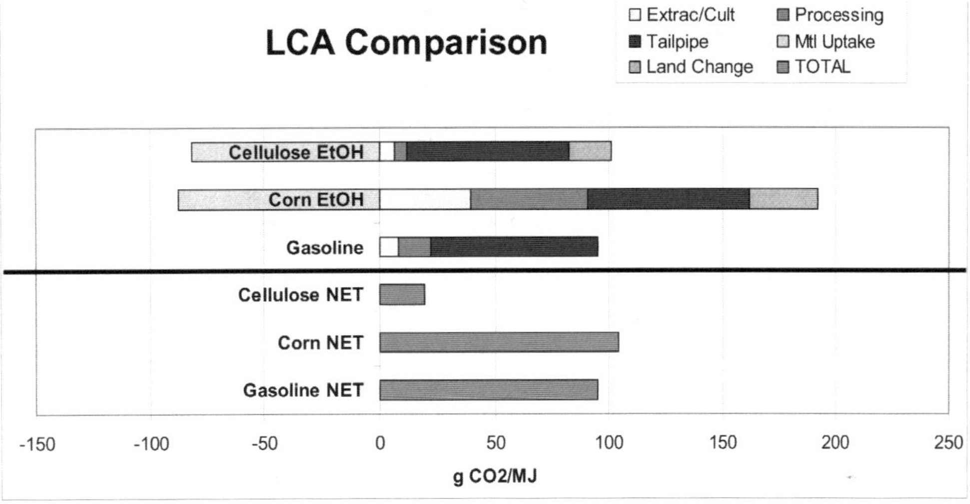

Data from CA Low Carbon Fuel Standard, 2009

Transportation Fuels Usage

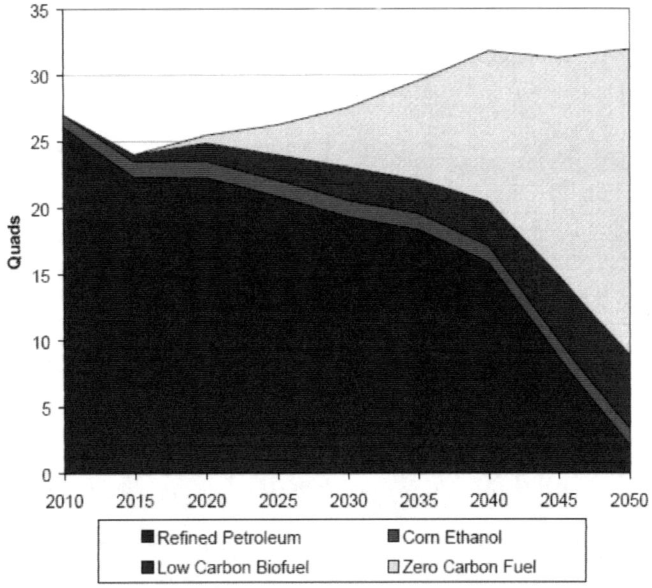

From CRA International Analysis of Lieberman-Warner Bill

MUSTANG

Refinery Case Studies

- 150 MBPSD Notional Refinery
- Product Slate
  - 8.4 MBPSD LPG
  - 14.8 MBPSD Jet
  - 78.8 MBPSD Gasoline
  - 43.0 MBPSD Diesel
  - 3.4 MBPSD No. 6 FO
  - 1,260 ton/day coke
- LCFS from California
- Cap-and Trade (Waxman-Markey)
  - EPA NPR for plant emissions
  - EIA for product emissions

MUSTANG

Refinery Case Study - LCFS

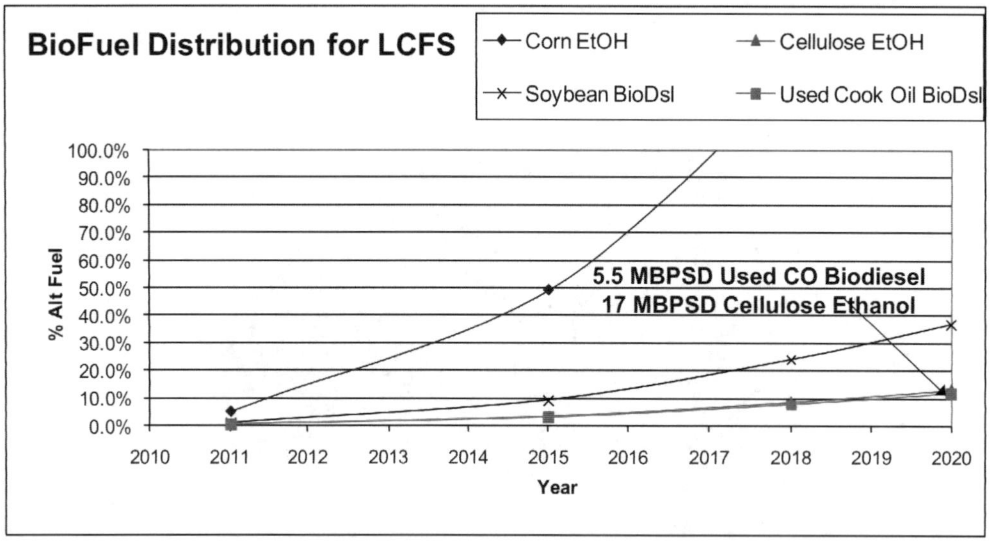

Refinery Case Study – Cap and Trade

- **Assumptions**
  - **Each case does not increase production over time**
    - **For bio fuels, equal replaced transportation products on an energy basis**
    - **LPG/Coke rates reduced to meet Gasoline reduction**
    - **CAFÉ impacts reduces gasoline and gasoline/distillate ratio**
  - **Market/Trading CO_2 prices taken from EIA evaluation of Waxman-Markey**
  - **10% allowances covered via offsets**
  - **Free allowances obtained accordingly**
    - **Refining Transition (2% till 2026)**
    - **Alternative fuel/CCS allowances (Obtained by refiner)**

Refinery Case Study – Cap and Trade

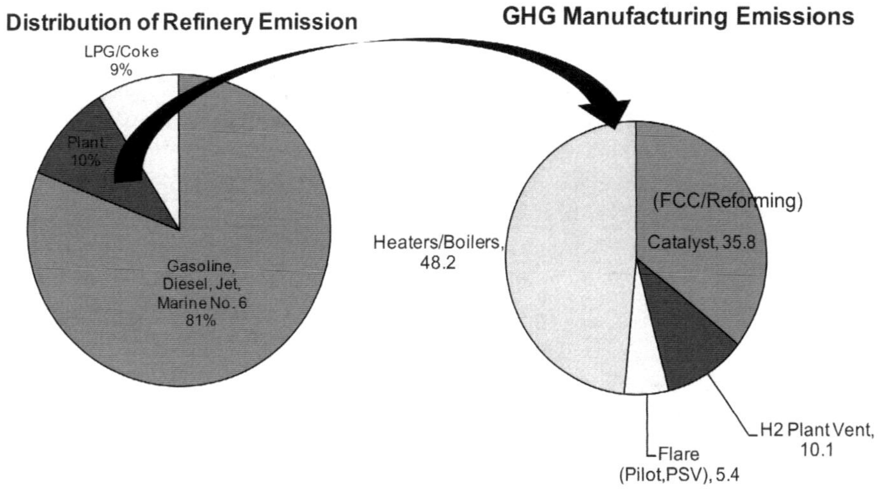

Refinery Case Study - Cap and Trade

- 2007 EIA and updated CAFÉ implications
- Assumptions
  - Car lifetime is 15 years
  - Even distribution of vehicle years on road
  - Fleet Average is the model year on-road average
  - 77% light duty vehicles use gasoline
  - Other mobile transportation CO_2 is constant
    - Jet/Marine/Off-Road/Heavy Duty
  - Net Light Duty miles driven is constant
  - Loss of energy for biofuels is accounted
    - Only Ethanol is considered

Refinery Case Study - Cap and Trade

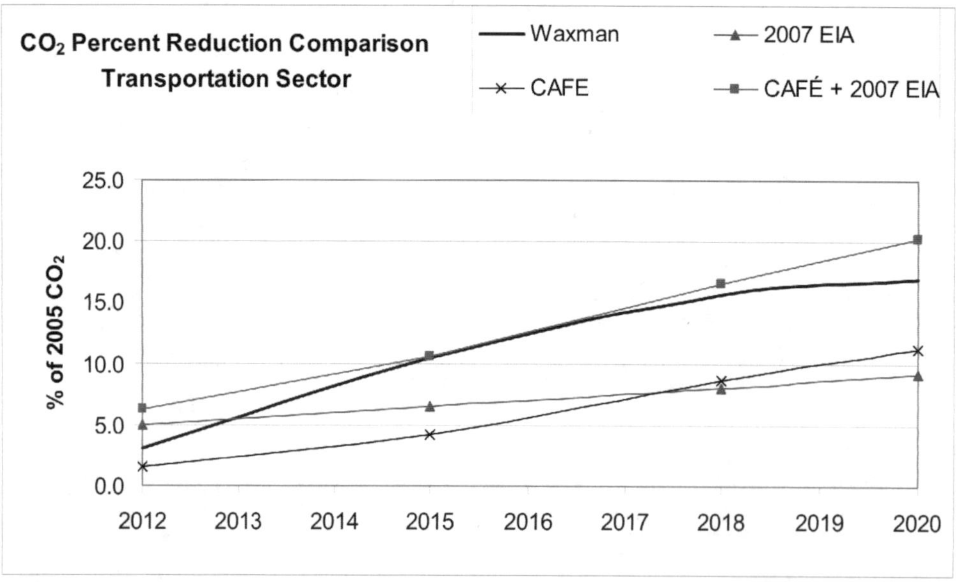

Refinery Case Study – Cap and Trade

MUSTANG

- Refinery H_2 demand likely to increase
  - Shift of Gasoline to Diesel products
    - RFS-2 and CAFÉ
  - Increased Clean Fuel Regulations
- Petroleum Coke Gasification
  - GHG emissions is refiner responsibility
  - Upgrades high Carbon material to H_2, steam, and electricity
  - Offsets H_2 Plant CO_2 emissions
  - Relatively easy to capture CO_2
  - Capital cost intensive

Refinery Case Studies – Cap and Trade

MUSTANG

- Refinery A
  - Do Nothing Base Case
- Refinery B
  - 10% Energy Efficiency Improvement in Manufacturing Emissions
- Refinery C
  - Petroleum Coke Gasification with CCS
- Refinery D
  - Include E10, BB15, B2, and J2 in product mix

Refinery Case Study – Cap and Trade

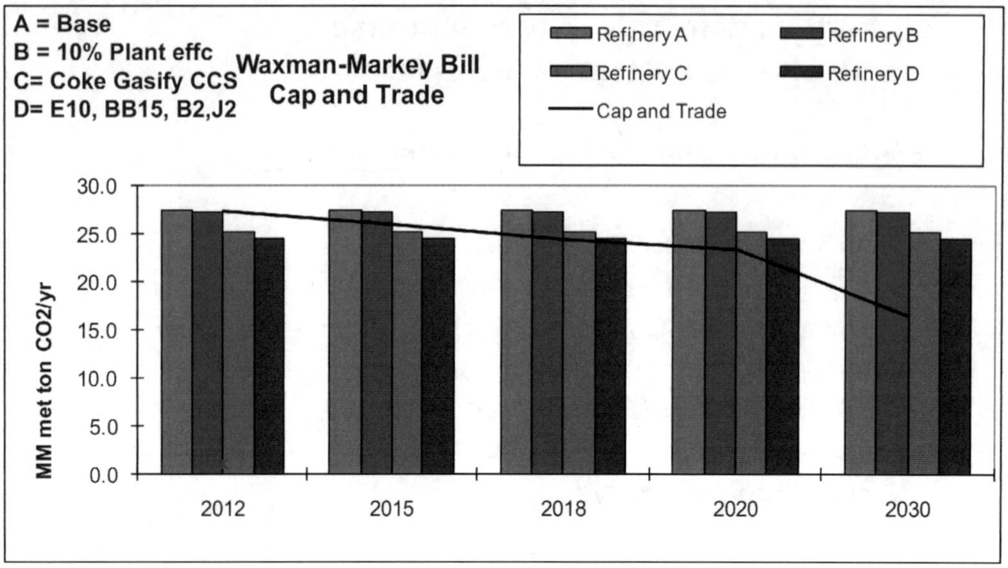

Refinery Case Study – Cap and Trade

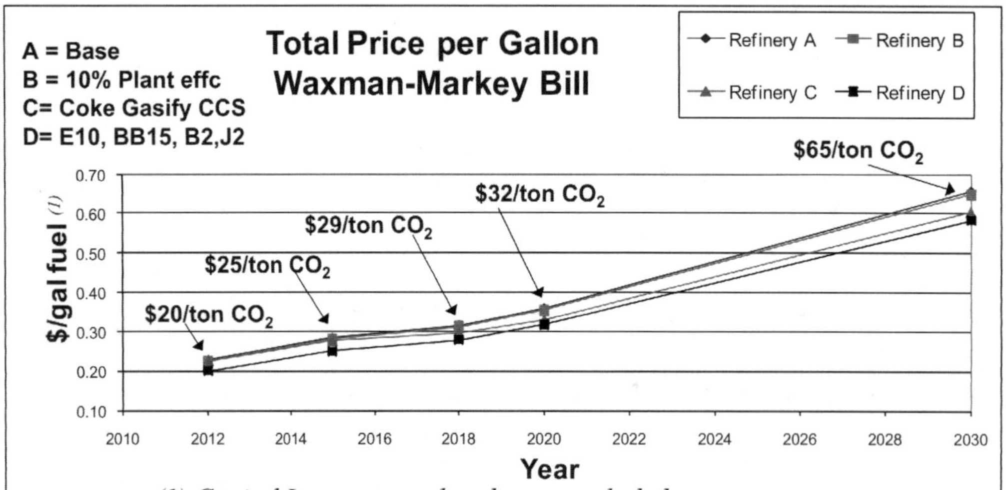

(1) Capital Investment related costs excluded

CO_2 price escalation from EIA Analysis "Energy Market and Economic Impacts of the American Clean Energy & Security Act '09"; "Basic Case" scenario utilized.

MUSTANG

Observations and Future Work

- With product emissions included in the cap-and-trade, refinery manufacturing emissions are only 10% of total responsibility
- Existing mandates (CAFÉ + 2007 EIA) that reduce transportation sector GHG emissions keep pace with Waxman-Markey to 2020
- Refinery emission reductions with Gasification of Petroleum Coke with CCS is comparable to current biofuel targets
- Identified Potential Solutions
  - Manufacturing Emissions
    - Crude Heater
    - Use of Heater Air Preheat
    - Targeted hot feeds and steam generation
    - Targeted hydraulic and expander utilization
  - Product Emissions
    - Integration of biofuels with existing liquid products
    - Gasification of Petroleum Coke with CCS

Crude Oil Compatibility for Mitigating Organic Fouling and Coking

Irwin (Irv) Wiehe
Soluble Solutions
www.SolubleSolutions.com

Current Issues and Opportunities in Refining

AIChE

Regional Process Technology Conference

Galveston **October 2, 2009**

Fouling Common in Crude Processing
Even Greater with Opportunity Crudes

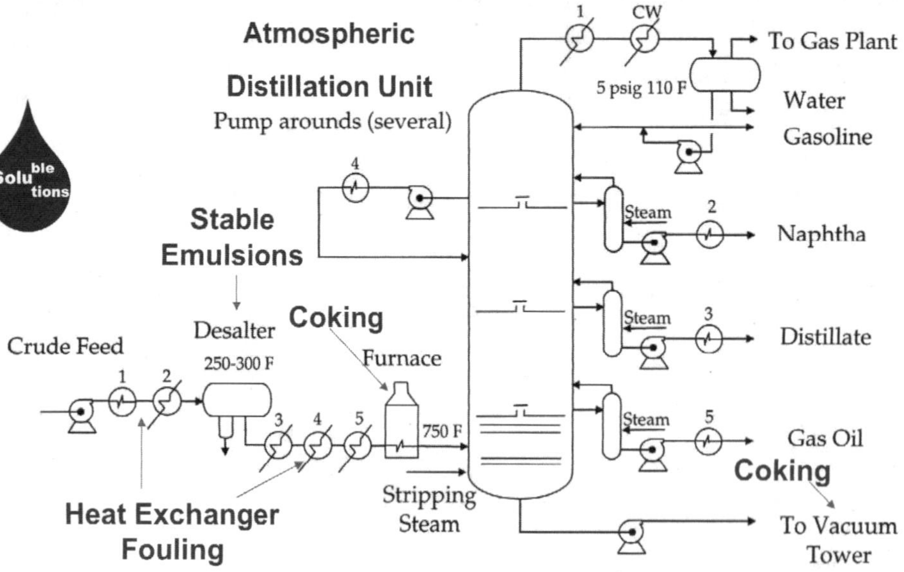

Organic Fouling - Crude Unit: Insoluble Asphaltenes

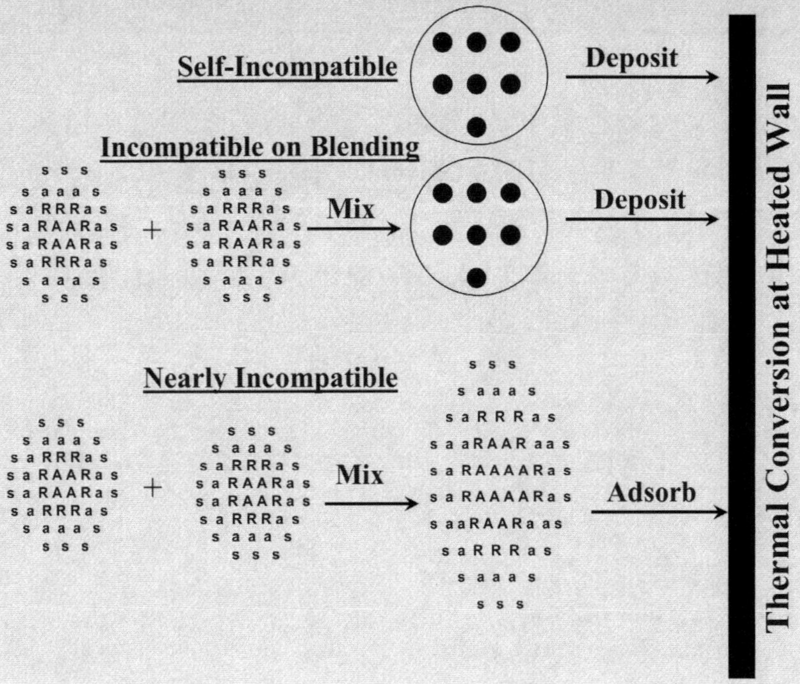

Oil Compatibility Model

- Measure Solvency (S_{BN}) and Insolubility (I_N) of Oils
  - Toluene: Model Solvent
  - n-Heptane: Model Nonsolvent
  - Oils Without Asphaltenes: Use Reference Oil
- Predict Compatibility and Incompatibility
  - Proportions of Blend
  - Order of Blending
  - Self-Incompatible
  - Nearly Incompatible
    - Fouling by Adsorption
    - Stable Oil-Water Emulsions

Oil Compatibility Model Has Unique Features

- Predicts Compatibility Operating Window for Oil Mixtures from Characterization of Oil Components
- Includes Both Oils that Contain Asphaltenes and Oils without Asphaltenes

Compatibility Criterion

Volumetric Average $S_{BN} > I_N$ of Each Component

Wiehe and Kennedy, Energy & Fuels, 14, 56-63 (2000).

Process for Blending Potentially Incompatible Oils
US Patent 5,871,634 (1999), Assigned to Exxon

Tests for Oil Compatibility Model (OCM)

- Blend Oil with Test Liquid (Toluene and n-Heptane)
- Determine Flocculation Points for Oil / Test Liquid Ratios

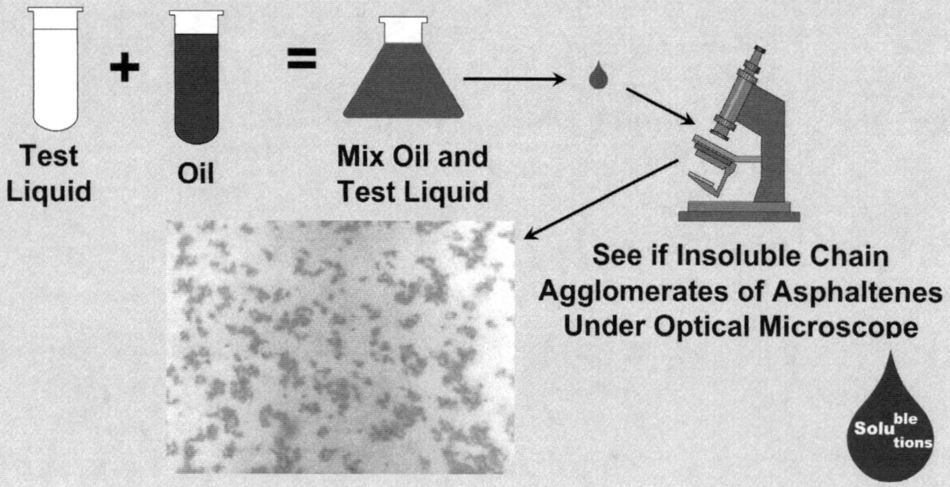

Data Verifies Linear Relation

Compatibility Numbers Calculated from Intercepts

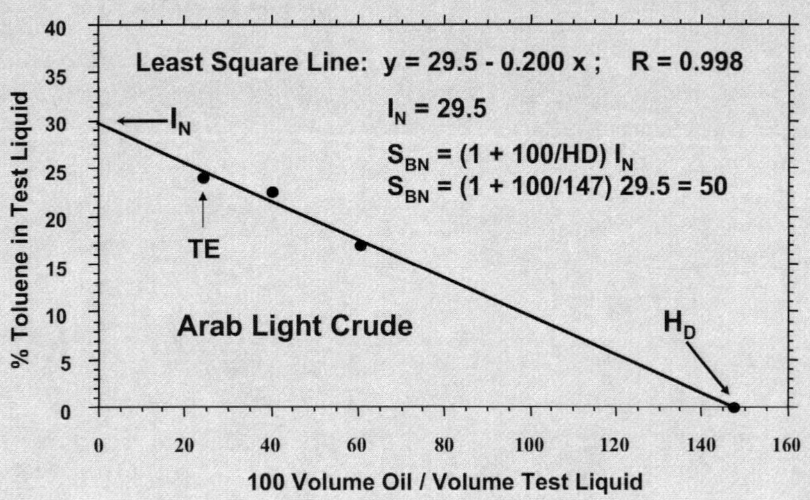

- Compatible Oils: $S_{BN} > I_N$
- Other Tests for Oils Without Asphaltenes

Tests for Oil Compatibility Model

Oils without Asphaltenes

- Select a Reference Oil that Contains Asphaltenes and TE and HD Tests Have Been Run
- Mix 5 ml Reference Oil and 25 ml Test Oil
  - If Compatible, Rerun TE Test on 2 g Reference Oil but Replace Toluene with Test Oil to Get SOE:

$$S_{BN} = 100\ TE_{REF} / SOE \qquad I_N = 0$$

  - If Incompatible, Rerun TE Test on 2 g Reference Oil but Replace n-Heptane with Test Oil to Get NSOE:

$$S_{BN} = \frac{100\ [TE_{REF} - NSOE]}{100\text{-}NSOE} \qquad I_N = 0$$

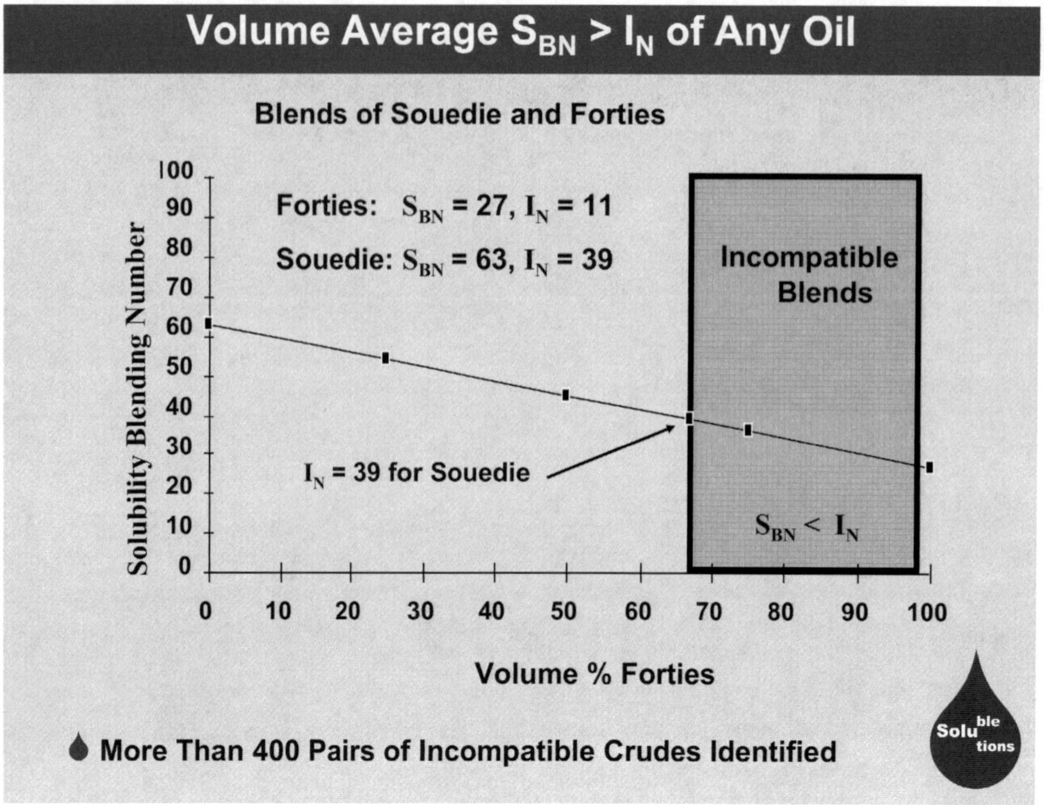
Volume Average S_{BN} > I_N of Any Oil
Blends of Souedie and Forties
Forties: S_{BN} = 27, I_N = 11
Souedie: S_{BN} = 63, I_N = 39
Incompatible Blends
I_N = 39 for Souedie
S_{BN} < I_N
Solubility Blending Number
Volume % Forties
More Than 400 Pairs of Incompatible Crudes Identified
Soluble tions

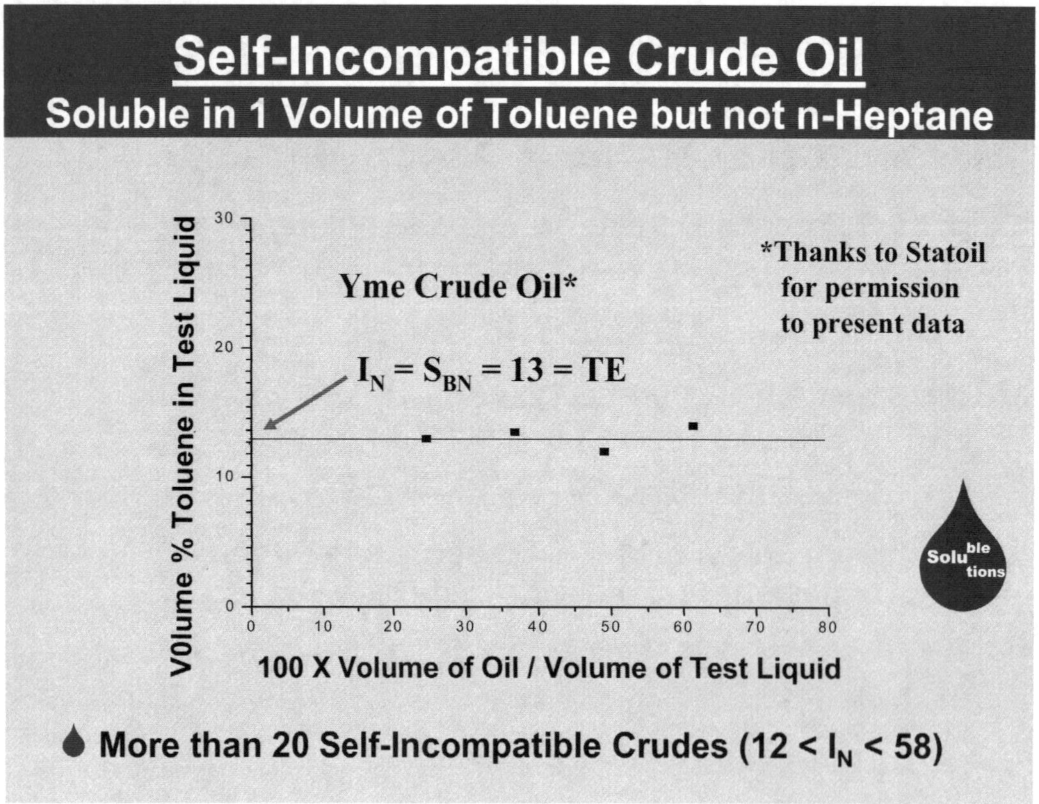
Self-Incompatible Crude Oil
Soluble in 1 Volume of Toluene but not n-Heptane
Yme Crude Oil*
I_N = S_{BN} = 13 = TE
*Thanks to Statoil for permission to present data
V0lume % Toluene in Test Liquid
100 X Volume of Oil / Volume of Test Liquid
More than 20 Self-Incompatible Crudes (12 < I_N < 58)
Soluble tions

Nearly Incompatible Oils are Moderately Fouling

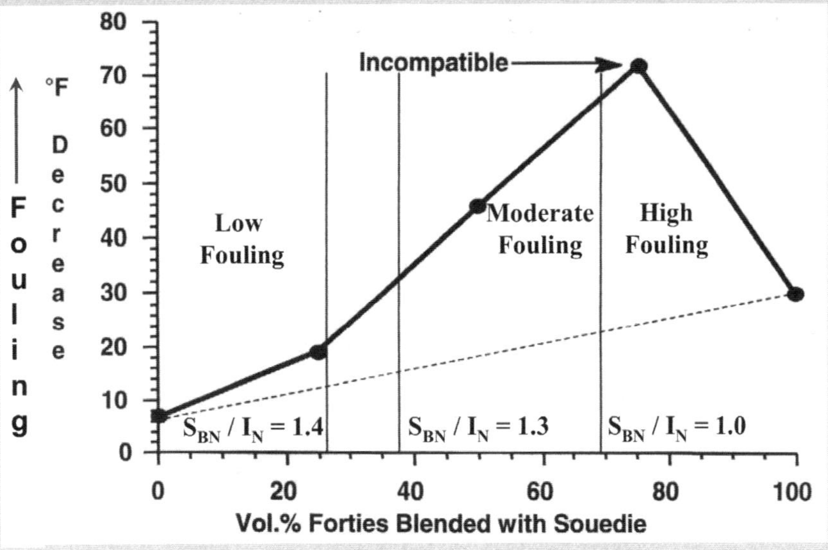

- Nearly Incompatible Crudes: One of the Causes of Desalter Rag Layers

Compatibility of Pairs of Crude Oils Is Greater than Predicted Without Accounting for Dispersion by Resins

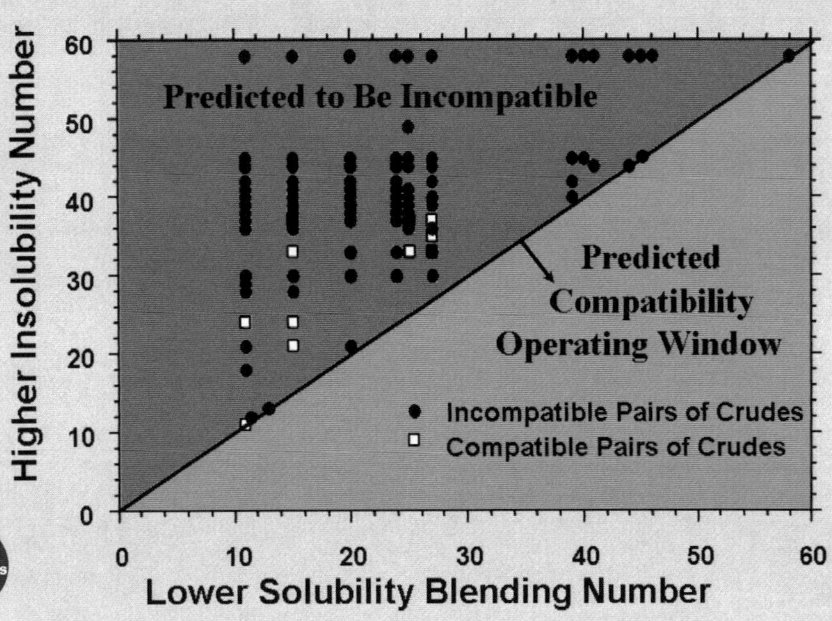

Best Practice for Crude Fouling Mitigation

Measure I_N & S_{BN} of Each Crude Oil of Interest

- Reject Purchase of Self- Incompatible Crudes
- Do Not Blend Crudes Incompatible in Any Proportion (I_N of One Crude > S_{BN} Another Crude)

Soluble Solutions Provides Testing Service

- More Details in Book, for 25% Off See www.SolubleSolutions.com

Soluble Solutions

*** Formatting Issue - Best Available Paper/Graphic ***

Options For Increasing Refinery Profitability

AIChE
Process Technology Conference

Outline

- Increased crude flexibility
- Improved crude selection
  - Updated crude assays
  - Increased planning LP accuracy
- Energy efficiency
- Petrochemical Integration

*** Formatting Issue - Best Available Paper/Graphic ***

Increased Crude Flexibilit

Increased Crude Oil Flexibility

- Enable purchase of most profitabl crude
- In the past this was heavier crude
- Currently light crude are more profitable

*** Formatting Issue - Best Available Paper/Graphic ***

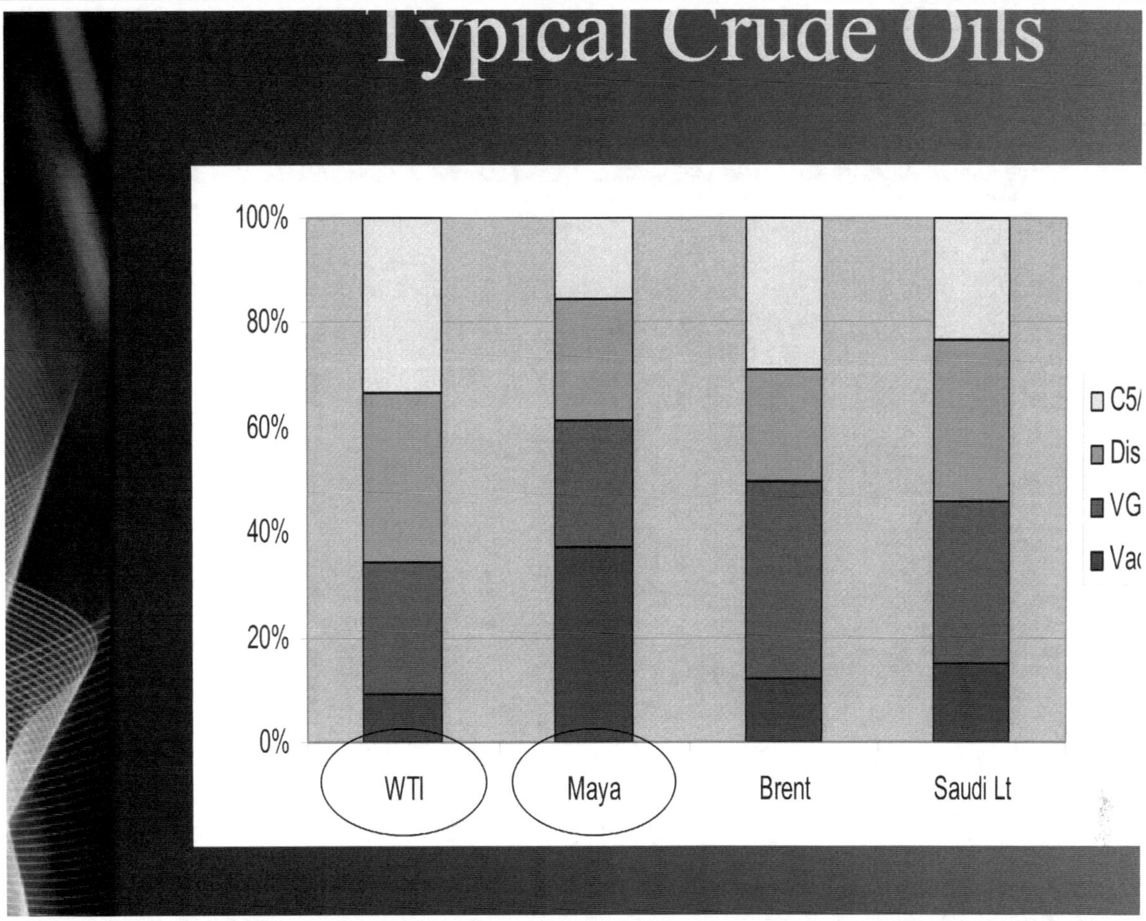

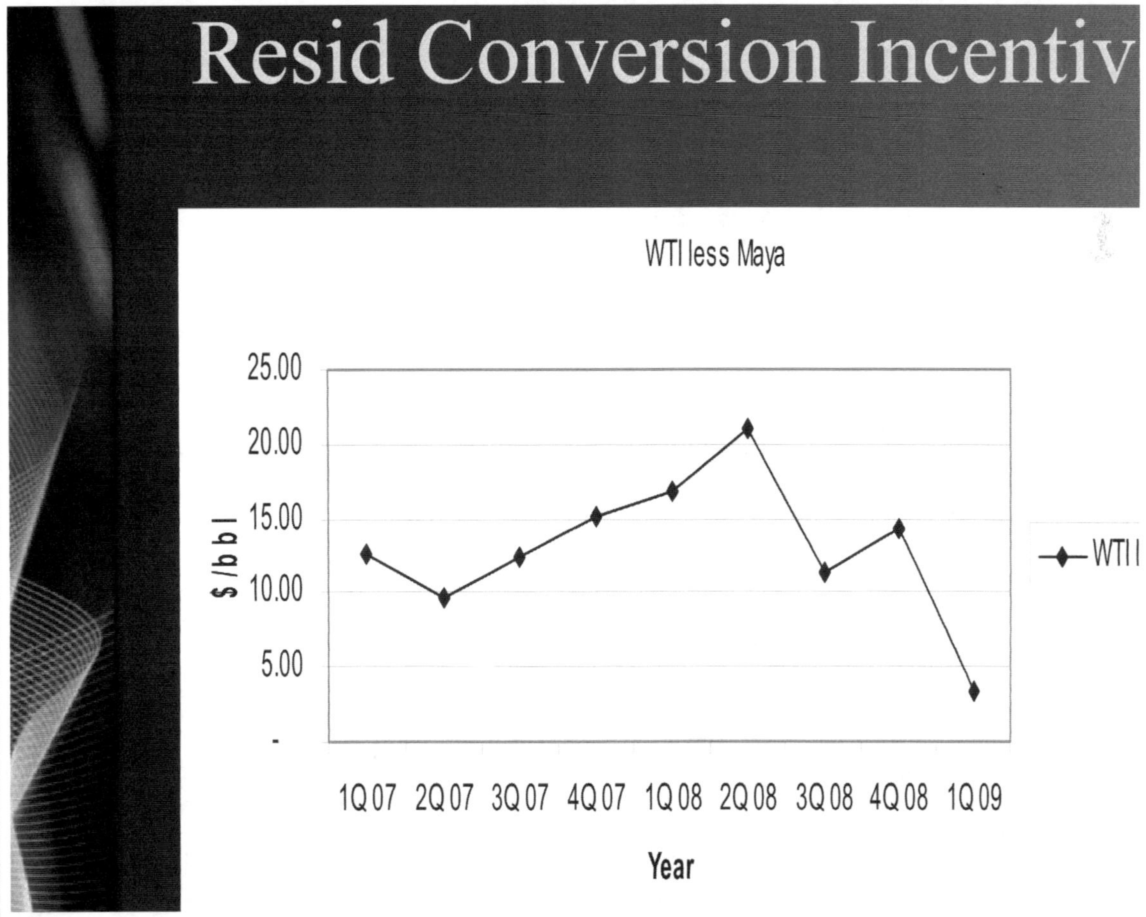

*** Formatting Issue - Best Available Paper/Graphic ***

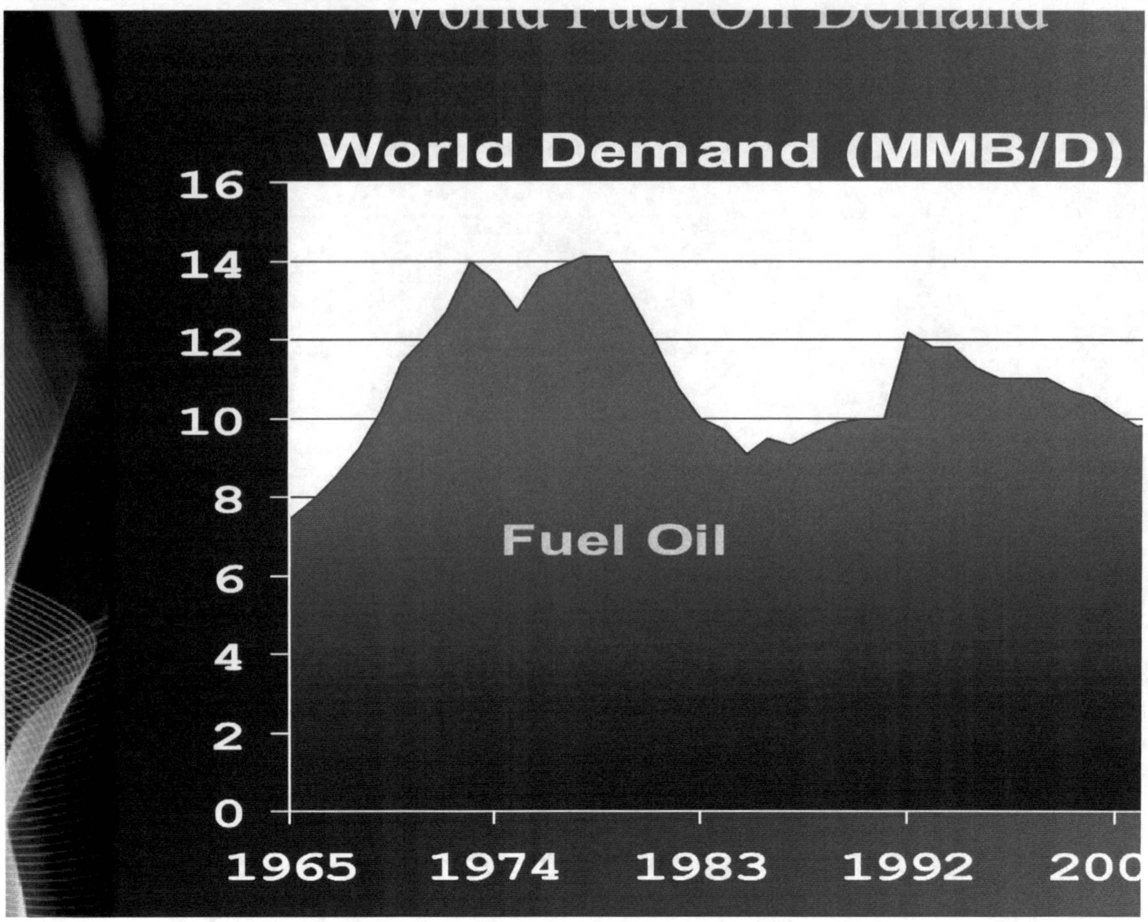

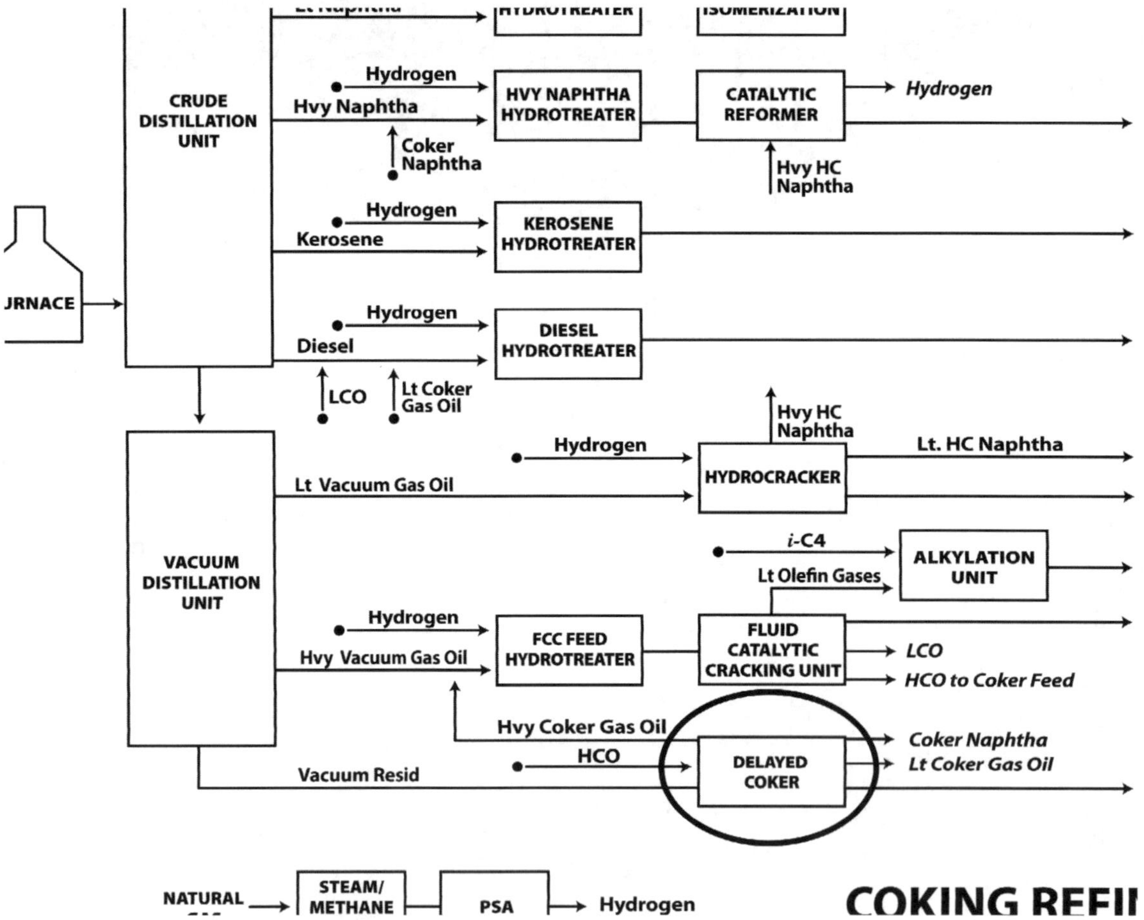

U. S. Coking Capacity

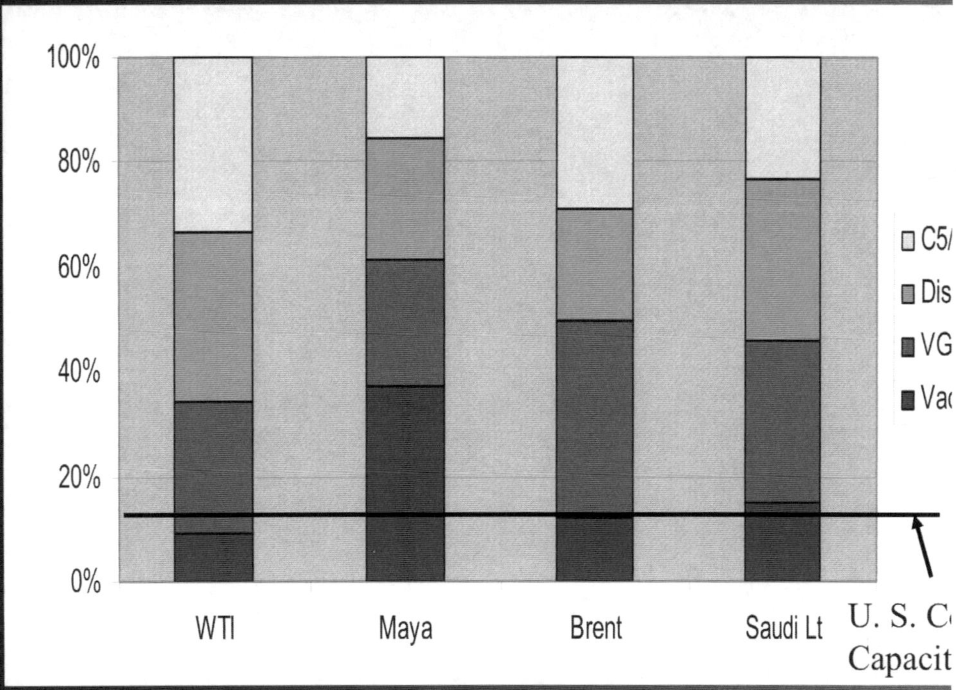

Delayed Coking-
Potential Projects

- New delayed coker
- Increase vacuum tower capacity and increase VGO/resid cutpoint
- Additional coke drum capacity
  - Replace existing drums with larger drums
  - Add additional drums
- Add upstream solvent deasphalter

Improved Crude Selection

Example Cost Structure

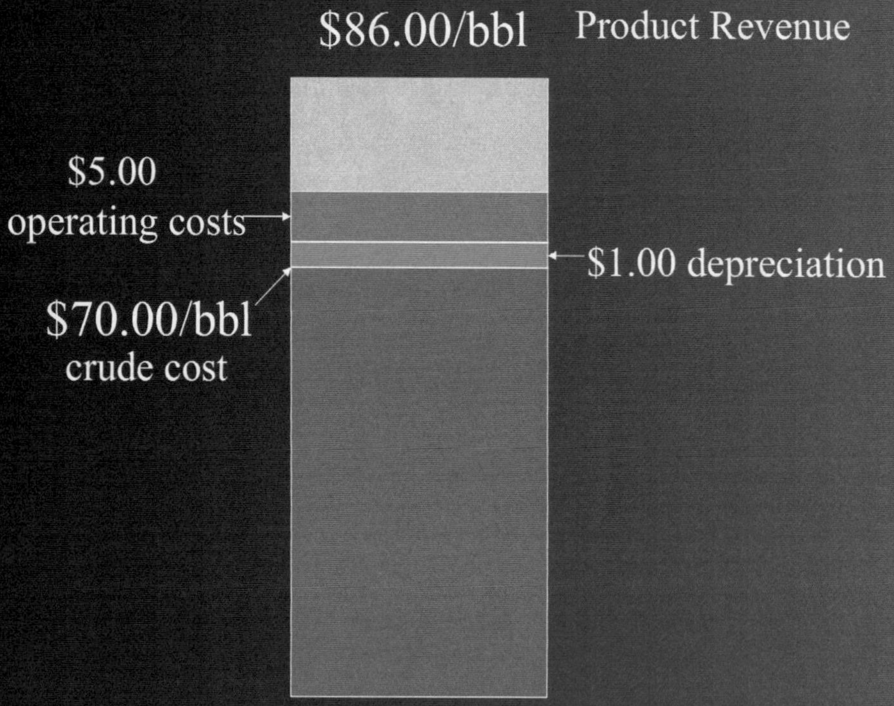

*** Formatting Issue - Best Available Paper/Graphic ***

Proper Crude Selection is V

- Typically 80-90% of total operating c
- Crude value is a function of many fac
  - **Crude oil assay**
  - Process units
  - Other crude oils being processed
  - Product demand and price

Improved Crude Selection

- Refinery LP used for crude selecti
- LP's require good data
- Accurate crude assays
  - Up-to-date crude assays
  - Valid crude evaluation procedure
- Accurate process models
  - Requires easy-to-use model updating system

*** Formatting Issue - Best Available Paper/Graphic ***

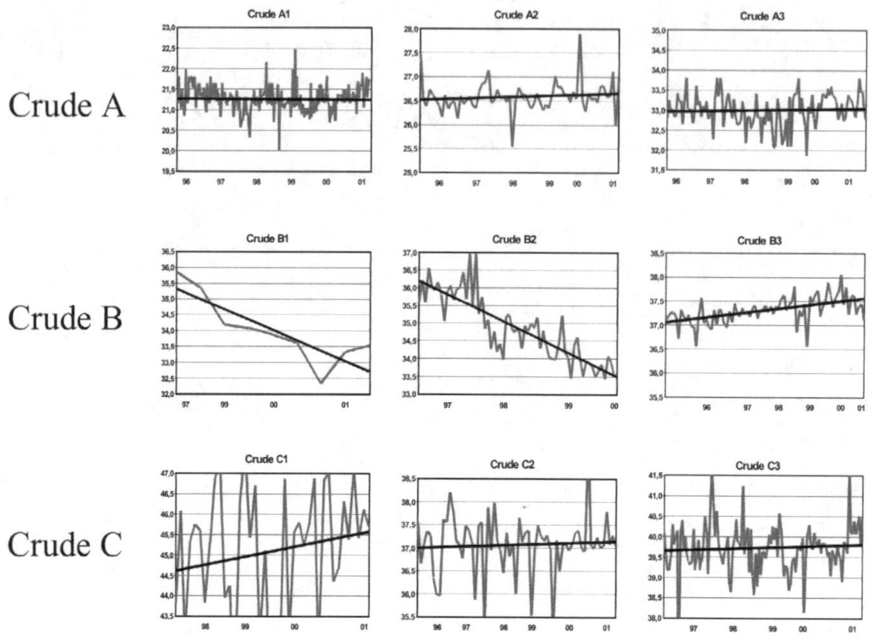

2

Crude Oil Assay System

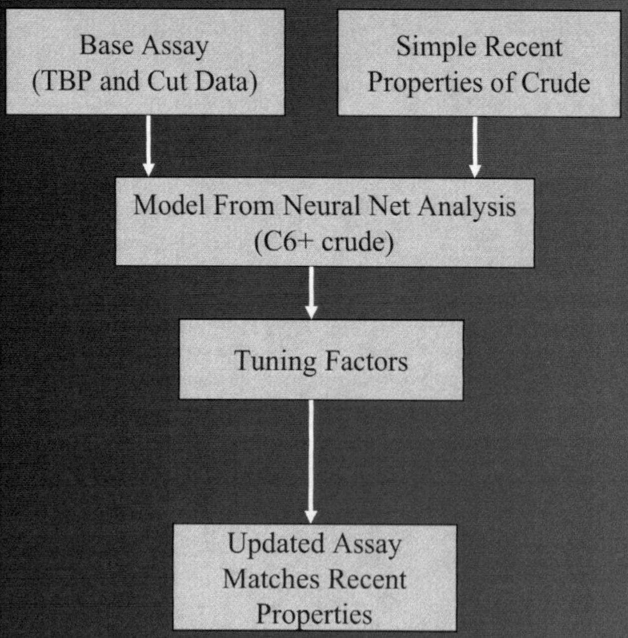

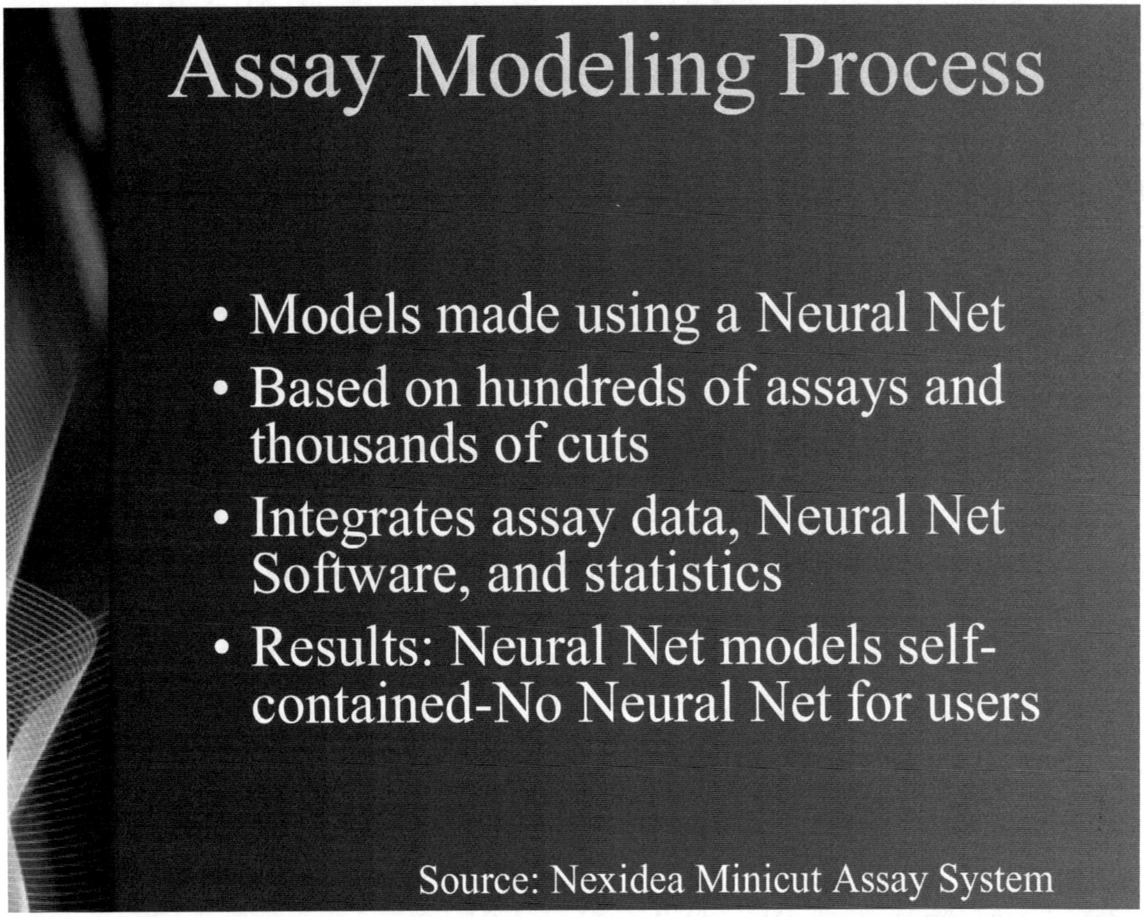
Assay Modeling Process
• Models made using a Neural Net
• Based on hundreds of assays and thousands of cuts
• Integrates assay data, Neural Net Software, and statistics
• Results: Neural Net models self-contained-No Neural Net for users
Source: Nexidea Minicut Assay System

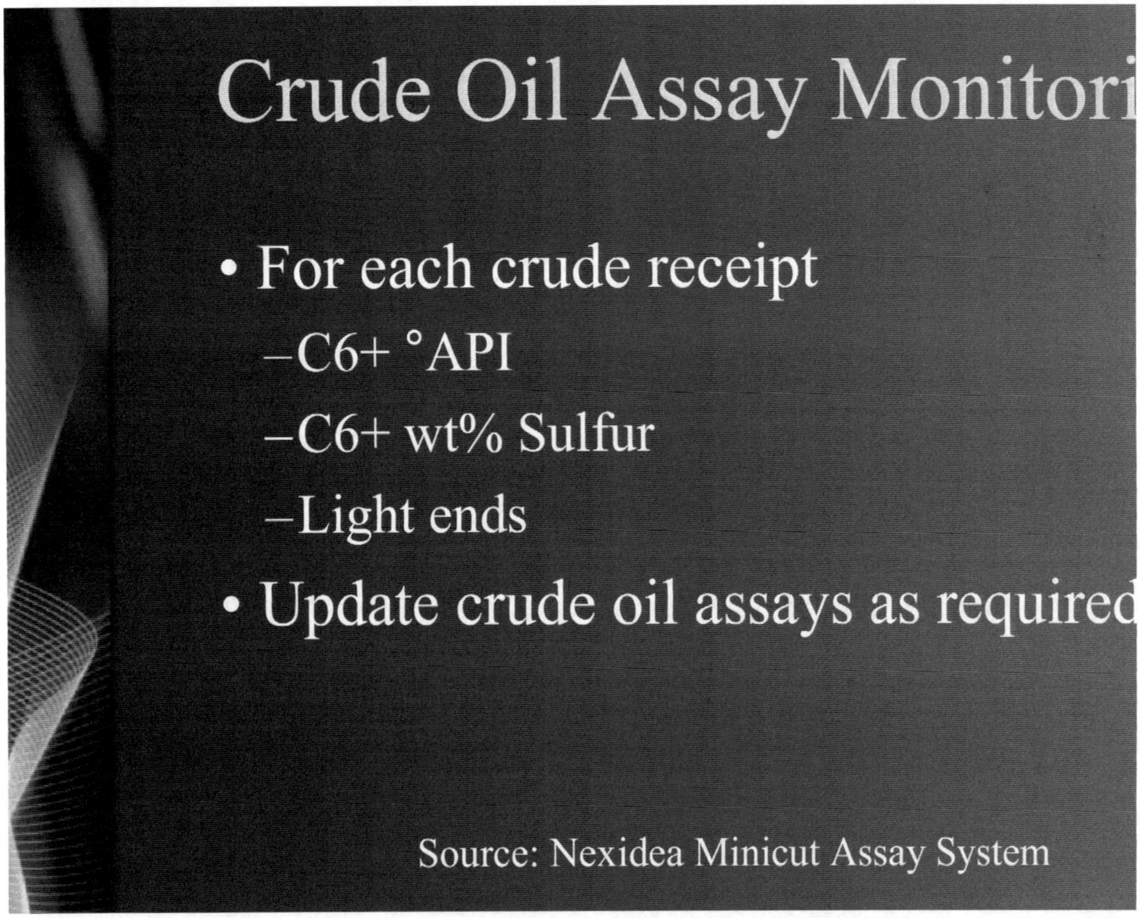
Crude Oil Assay Monitori
• For each crude receipt
–C6+ °API
–C6+ wt% Sulfur
–Light ends
• Update crude oil assays as require
Source: Nexidea Minicut Assay System

Crude Oil Assay Update

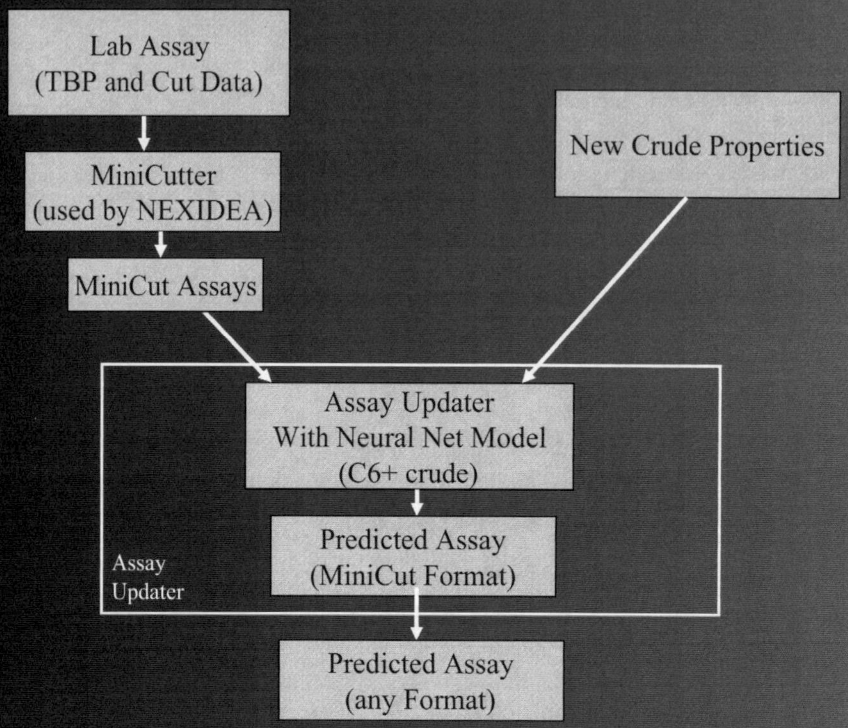

Updated Assays

- Updates TBP and cut properties
- Prepares new crude assays
- Recuts crude assays
- Allows better feedstock selection and production plans

Source: Nexidea Minicut Assay System

Accurate Process Models

LP Process Models

- Linear Models
  - Alkylation
  - Isomerization
- Non-linear Models
  - FCC
  - Hydrocracking
  - Delayed Coking

*** Formatting Issue - Best Available Paper/Graphic ***

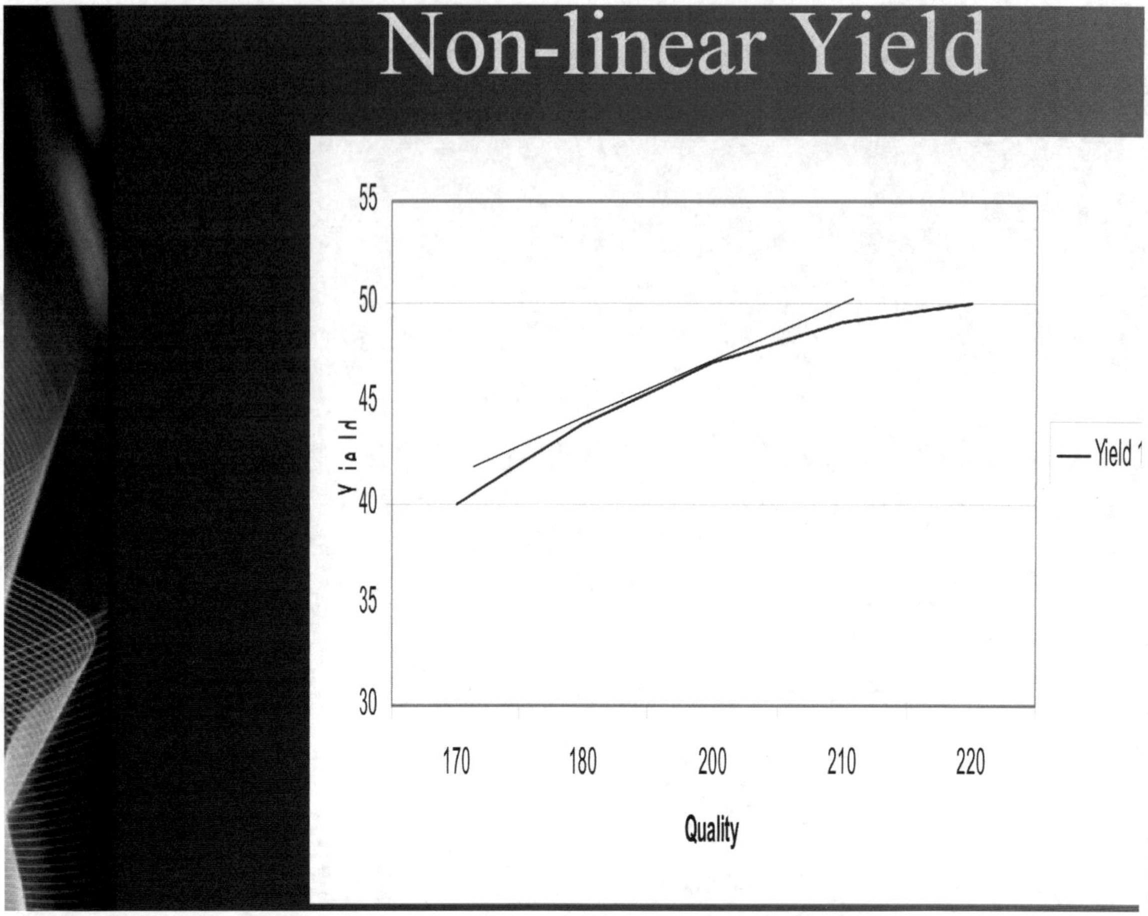

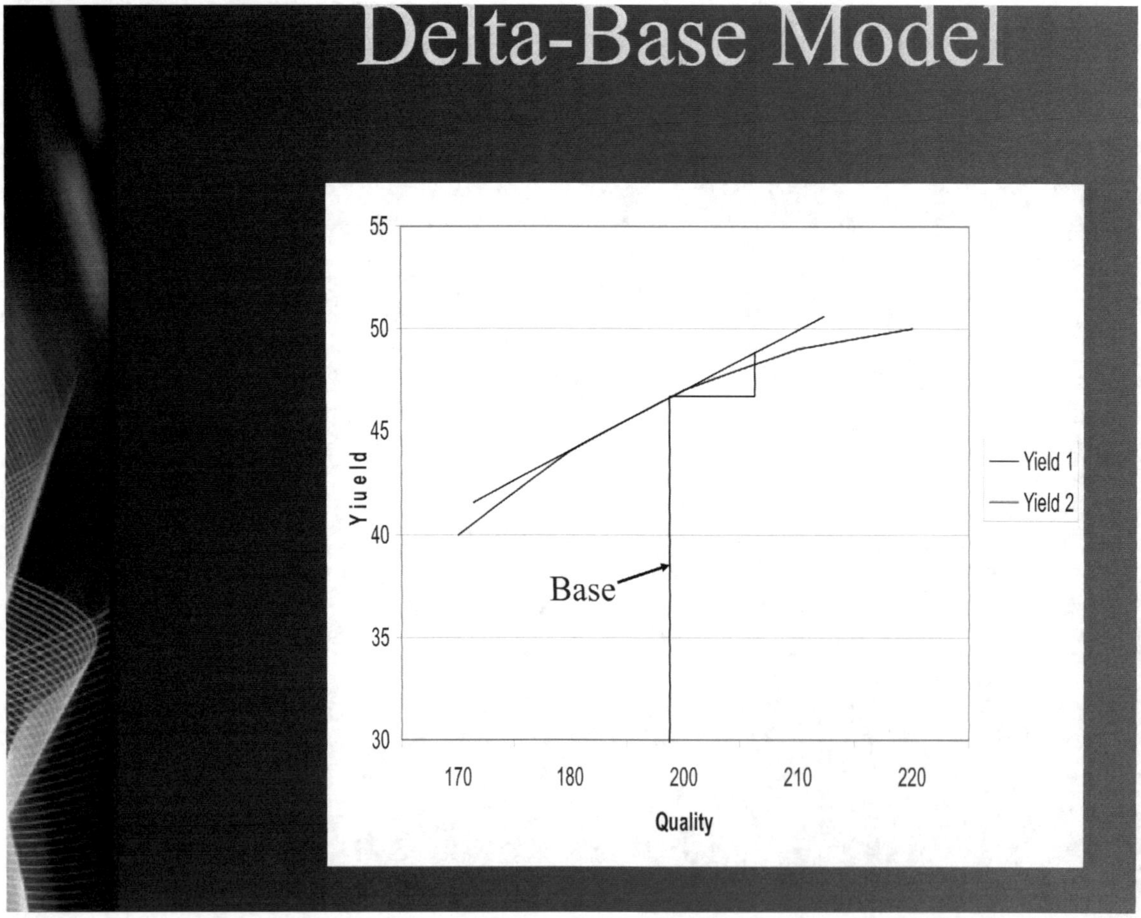

*** Formatting Issue - Best Available Paper/Graphic ***

Delta-Base Model Structure

	Feed	Base Yield	Delta
Feed	1.0	-1.0	
C5-180		16.2	4.4
180-400		40.0	10.0
API	22	-20.0	-2.0

Yields Change Over Time

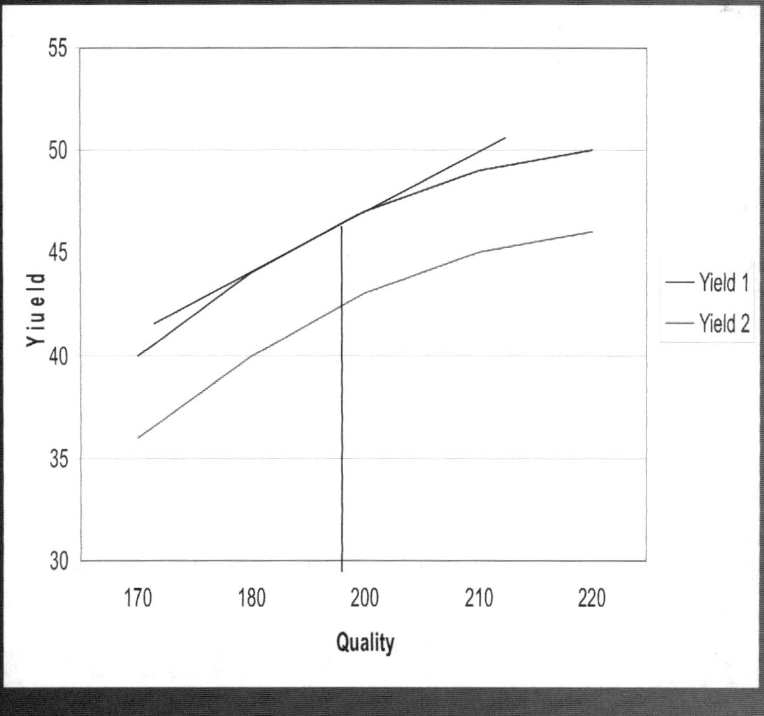

*** Formatting Issue - Best Available Paper/Graphic ***

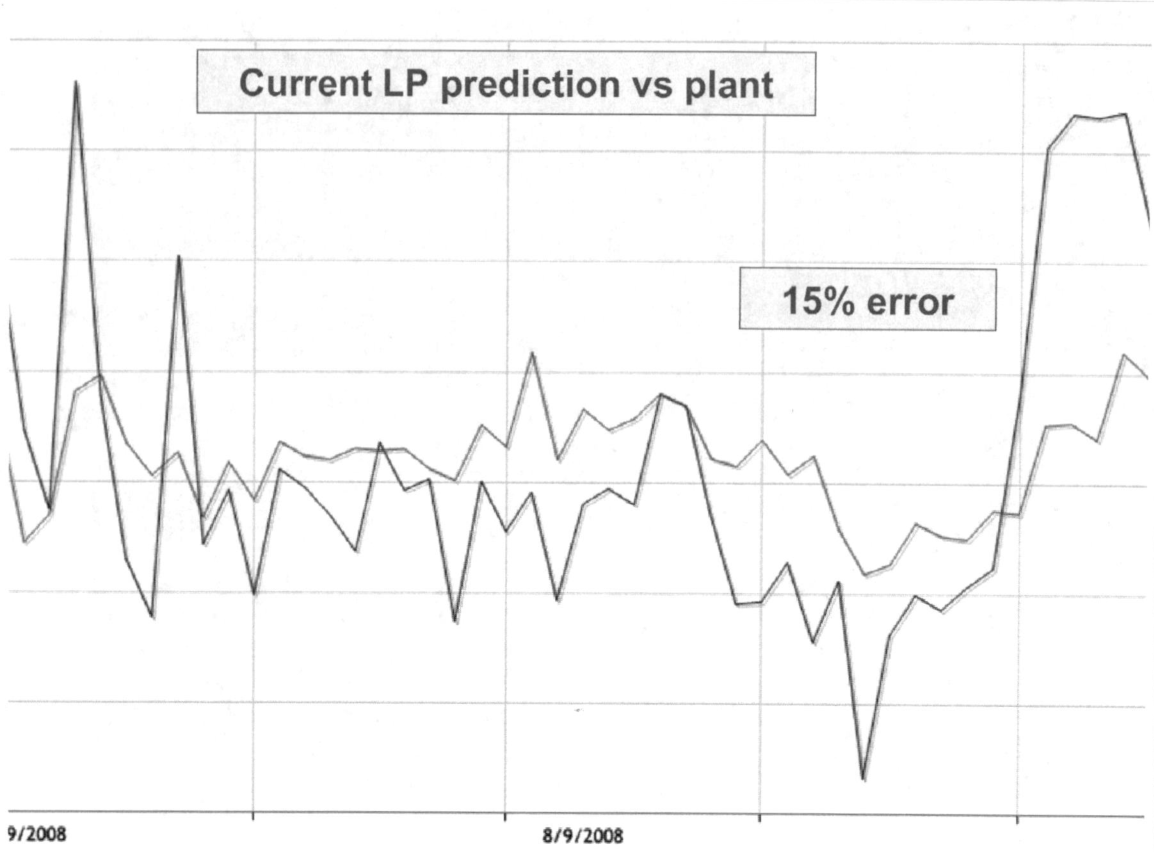

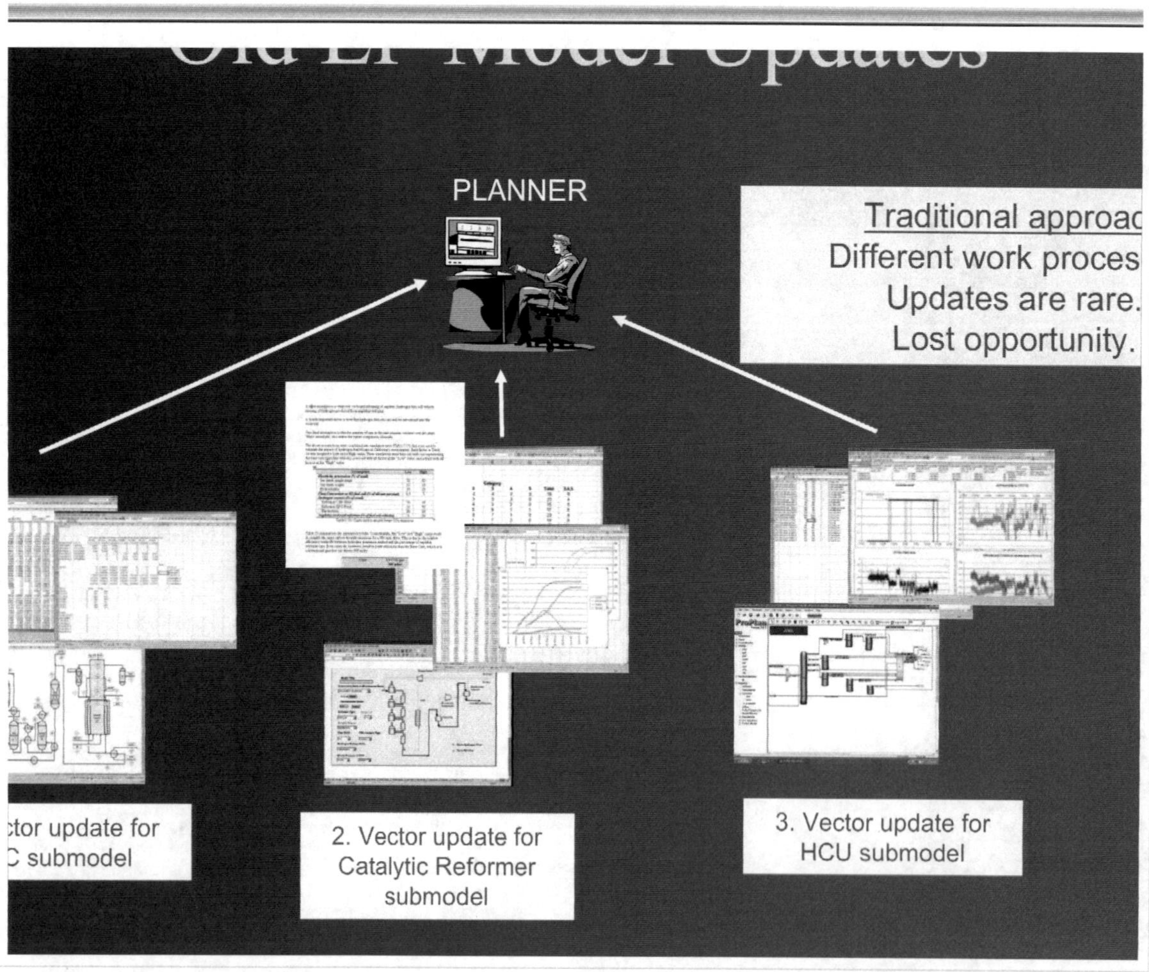

***** Formatting Issue - Best Available Paper/Graphic *****

Accurate Non-linear mode

- Compare LP predictions with the pl
  - Locate Planning LP errors.
- Improve LP model
  - Calculate new LP vectors based on plan data.
  - Test new LP vectors against plant or simulation data.
  - Implement system to export vectors directly into LP.

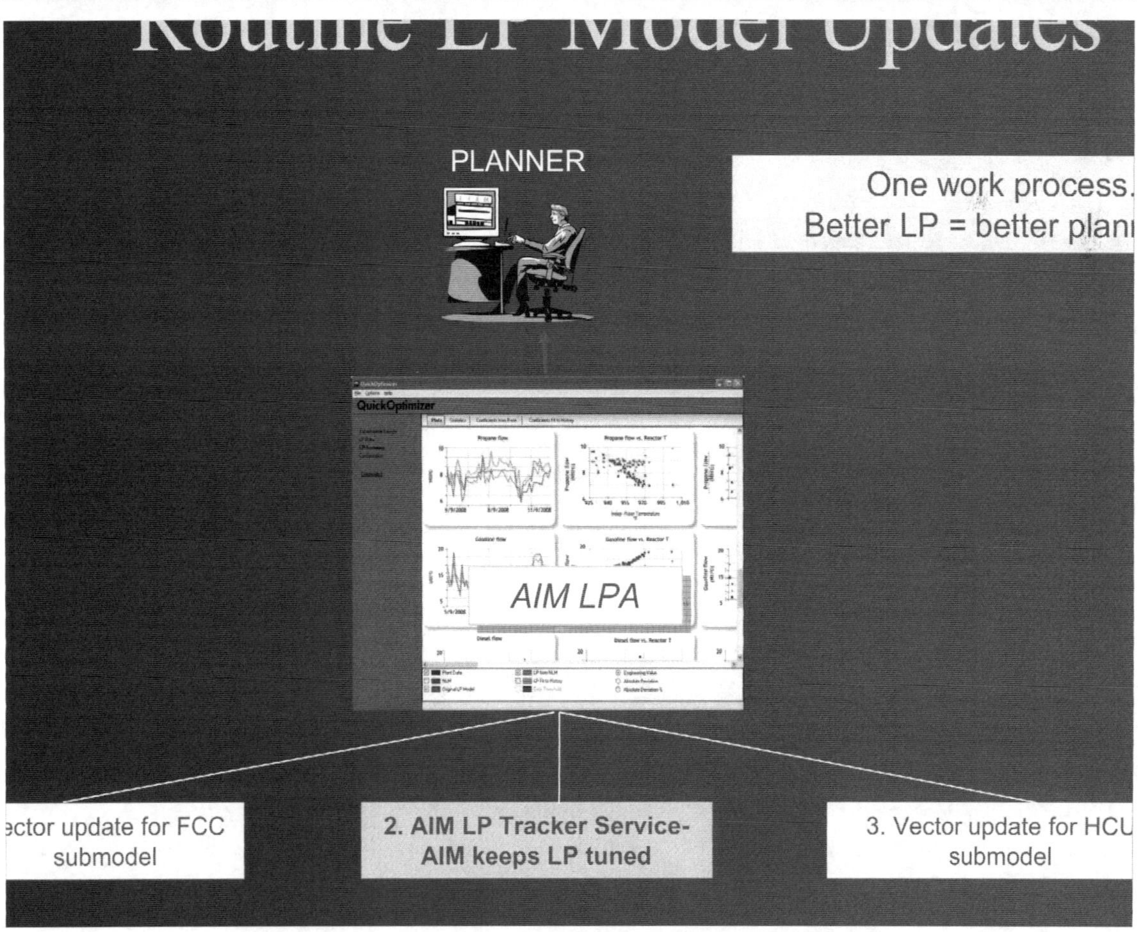

*** Formatting Issue - Best Available Paper/Graphic ***

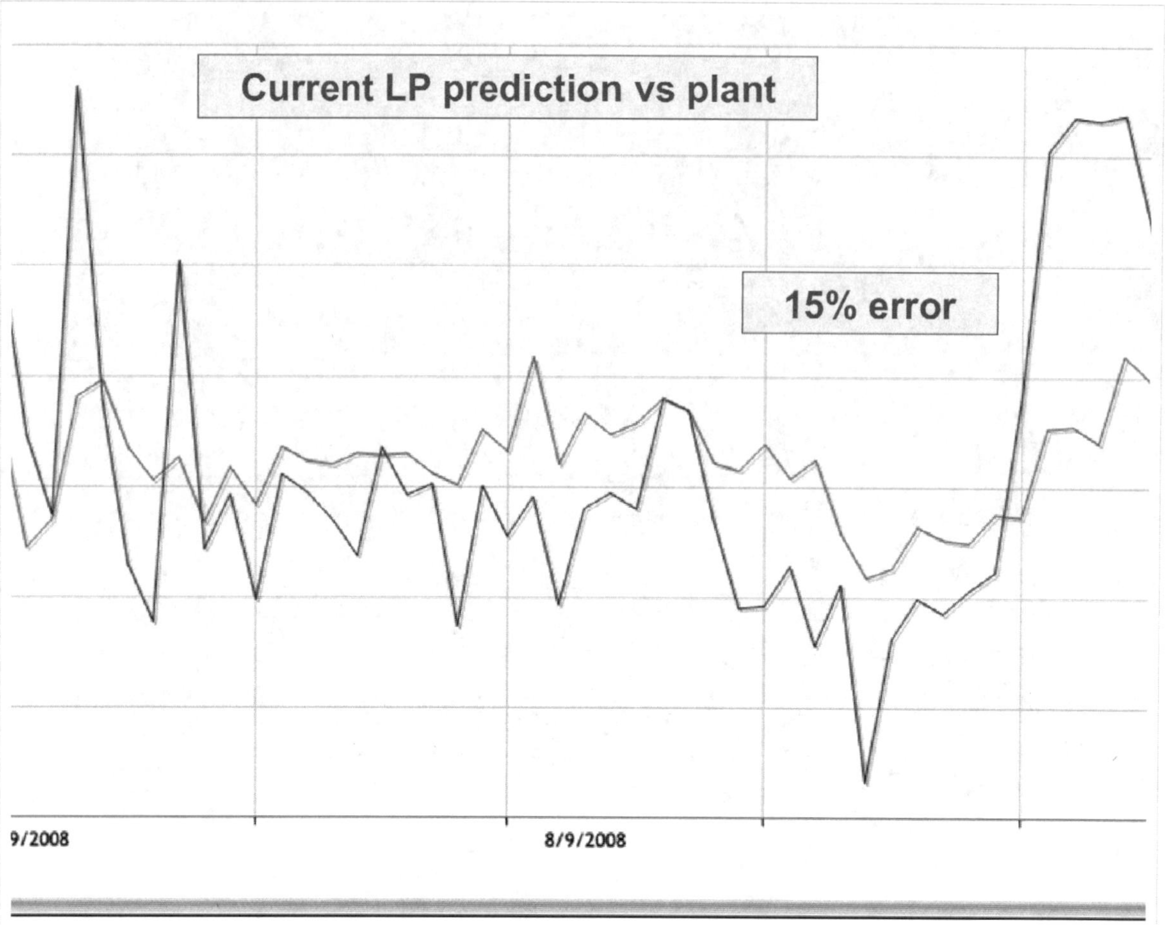

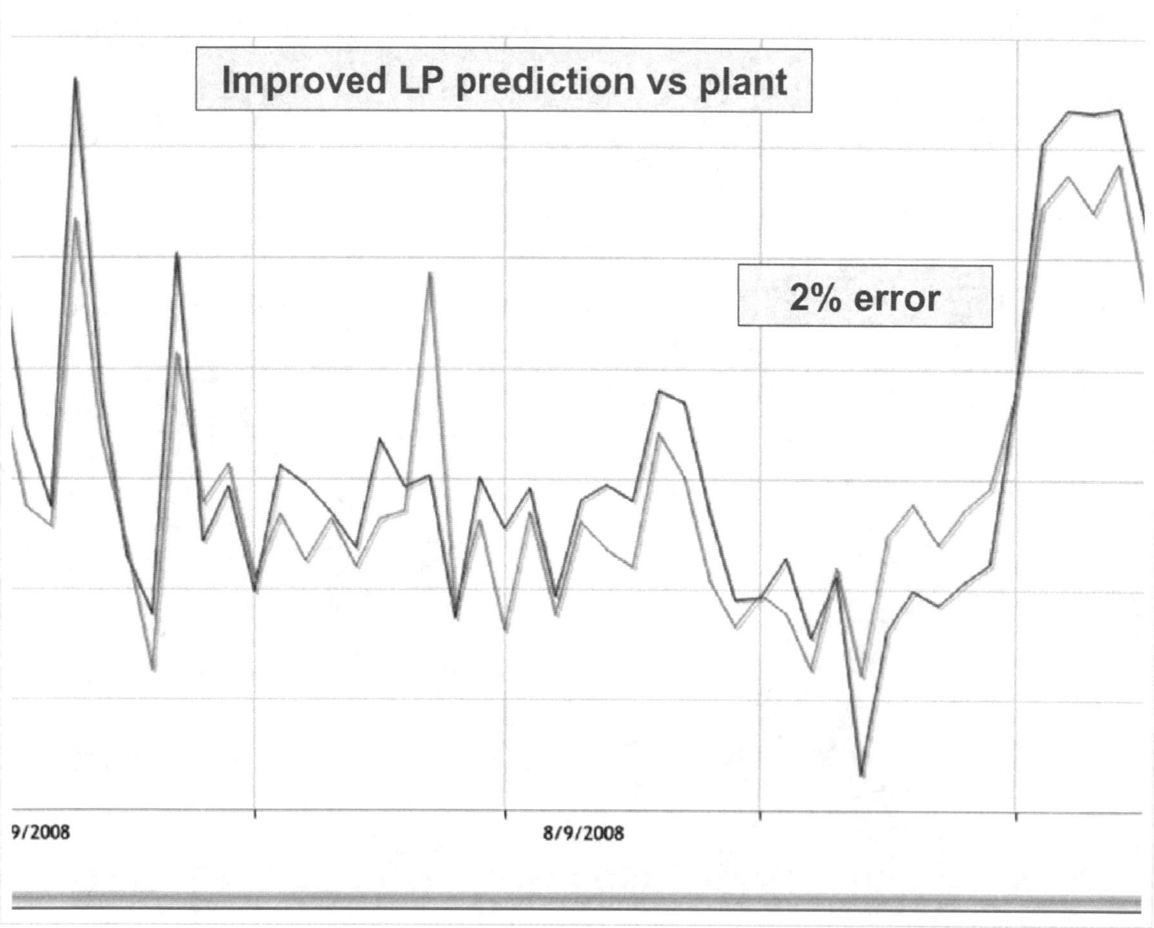

Workflow

- Export LP submodel spreadsheet to LP Assurance tool
- Automatically collect plant data
- Validate plant data
- Measure LP submodel accuracy
- Calculate new submodel coefficients
- Validate accuracy of new coefficients
- Export new coefficients to LP system

System Features

- Can be easily used by Planners
- Has automated data handling & screening
- Has a single used interface for Planners & Engineers
- Quickly calculated updated base coefficients

Summary

- Quick project payback
- Improved crude selection
- Improved production planning
- Increased LP model credibility – bett
match between plan vs actual.
- Retains staff expertise

Energy Efficiency

*** Formatting Issue - Best Available Paper/Graphic ***

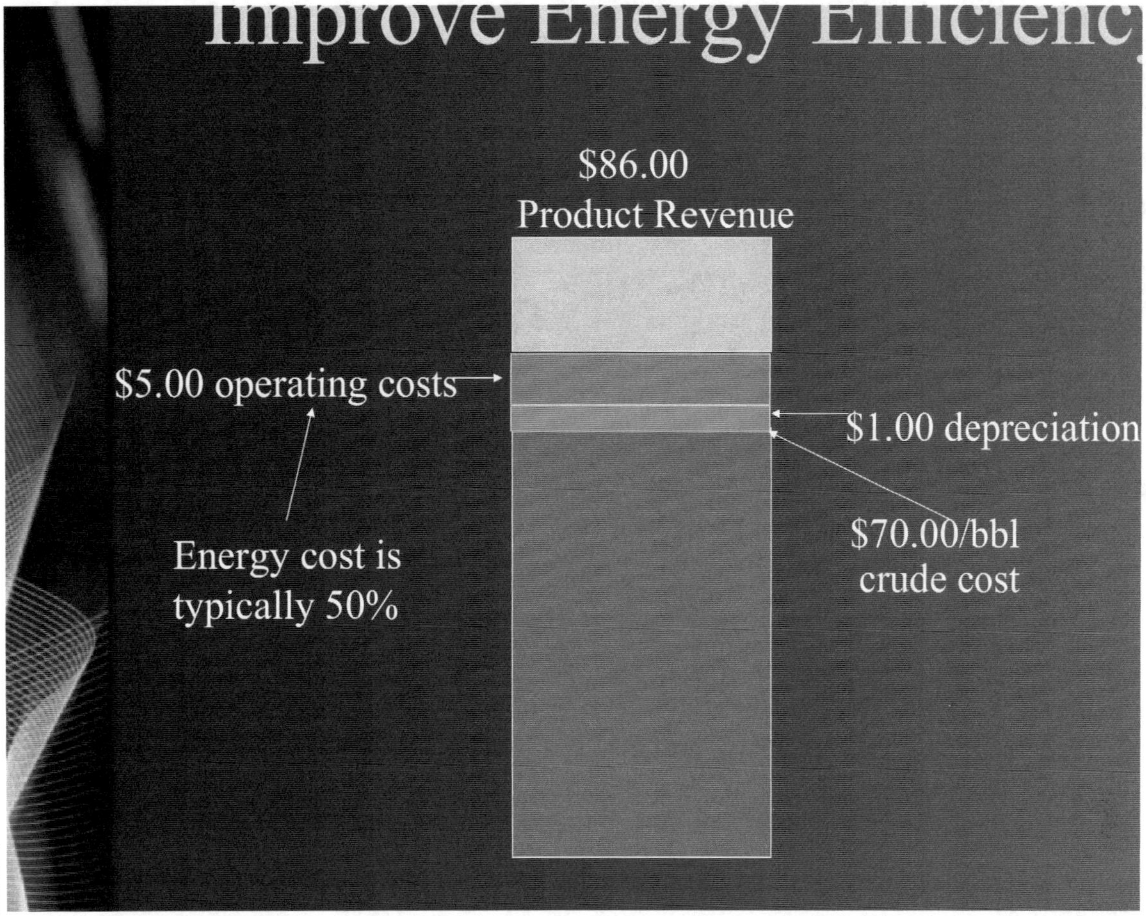

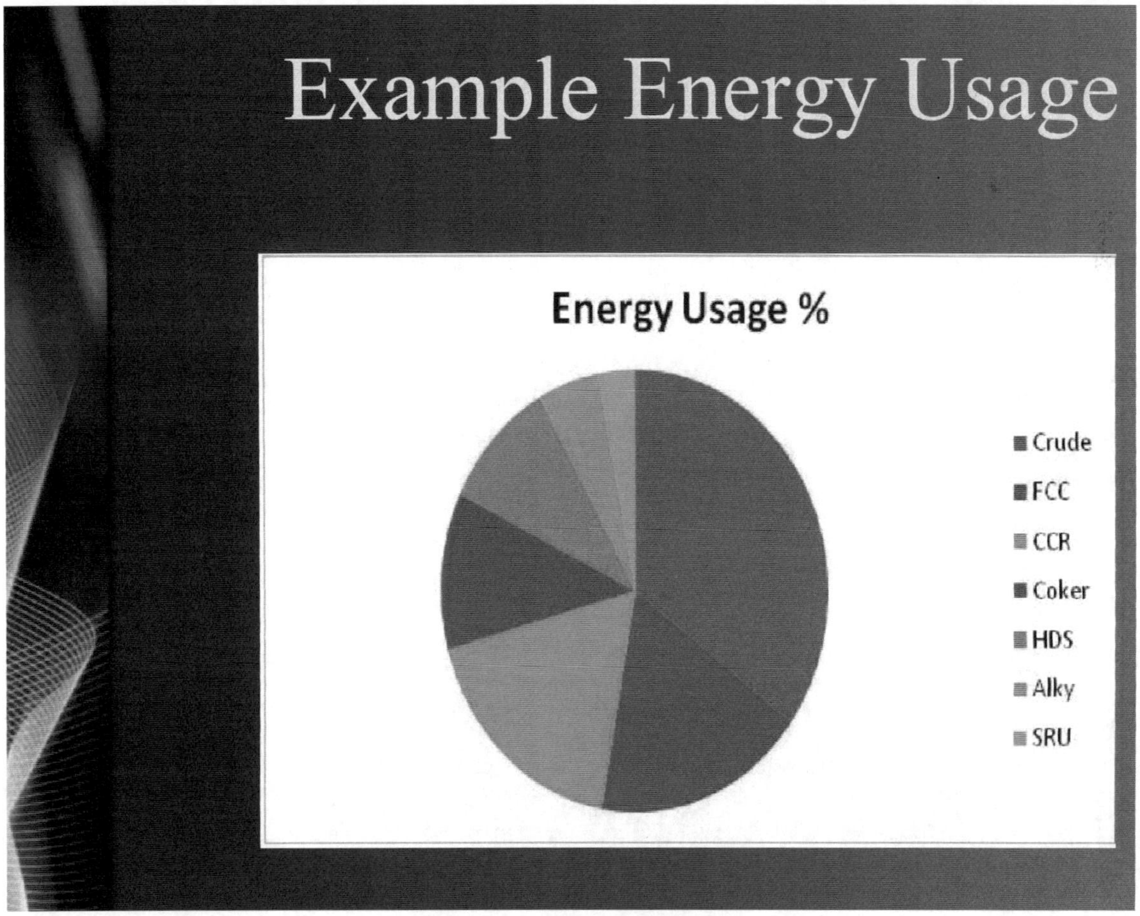

Example Energy Costs

- In 2008 energy costs for a 150,000bpd
  - $210,000,000/year
  - 58% of total operating costs
  - 41% produced by refinery
  - 17% purchased

Energy Efficiency Exampl

- Reduce furnace/boiler O_2
- Clean process preheat exchangers
- Utility system optimization
- Establish fuel system strategy

Process Areas

- Evaluate high energy processes
- Improvement areas
  - Reduce furnace temperatures
  - Reduce column pressures
- Establish dynamic energy targets
- Process energy consumption varies
  - Feedstock quality
  - Catalyst activity, etc.
- Implement monitoring system

Potential Projects

- Heat exchange train pinch study
- Cogeneration project to utilize low l heat
- Heater air preheat
- Waste heat boilers

Requires Long-term Committment

- Establish full-time energy manager
- Frequent recalculation of fuel system economics with updated prices
- Frequent recalculation of steam/power system economics
- Routinely monitor
  - Furnace excess air
  - Preheat exchanger train
  - Process energy efficiency

Petrochemical Integration

Aromatics

BTX-Gasoline Crack Comparison

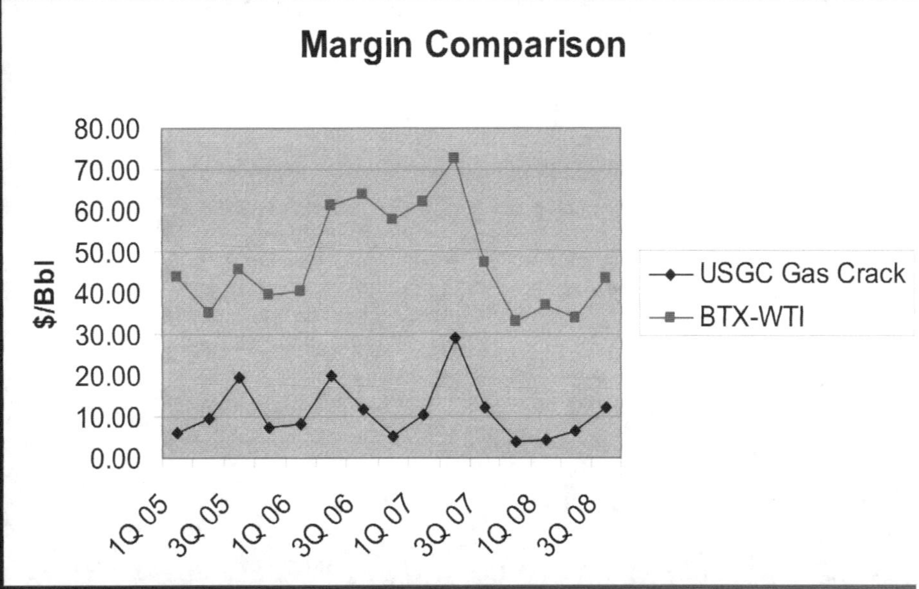

Integrated Aromatics Comp
Example Overall Yields

• Paraxylene	45%
• Benzene	15
• Orthoxylene	10

Olefins Production

Olefins Production

- Feedstocks
  - Ethane
  - Propane
  - Naphtha, etc.
- Major Products
  - Ethylene
  - Propylene

Refinery/Petrochemical Integration

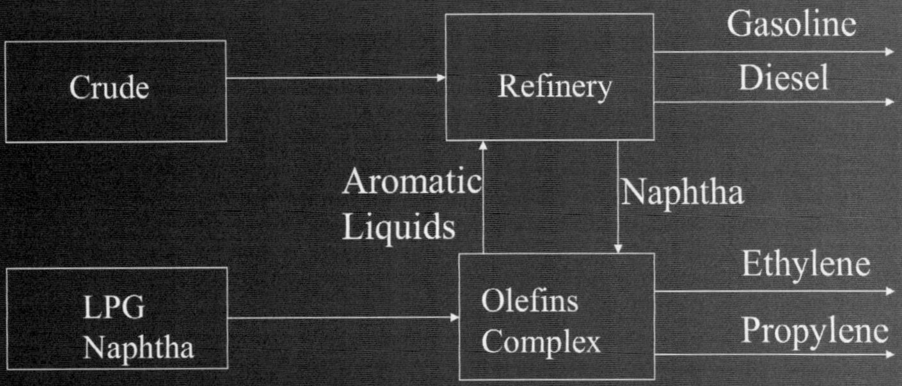

*** Formatting Issue - Best Available Paper/Graphic ***

Site-Wide Integration

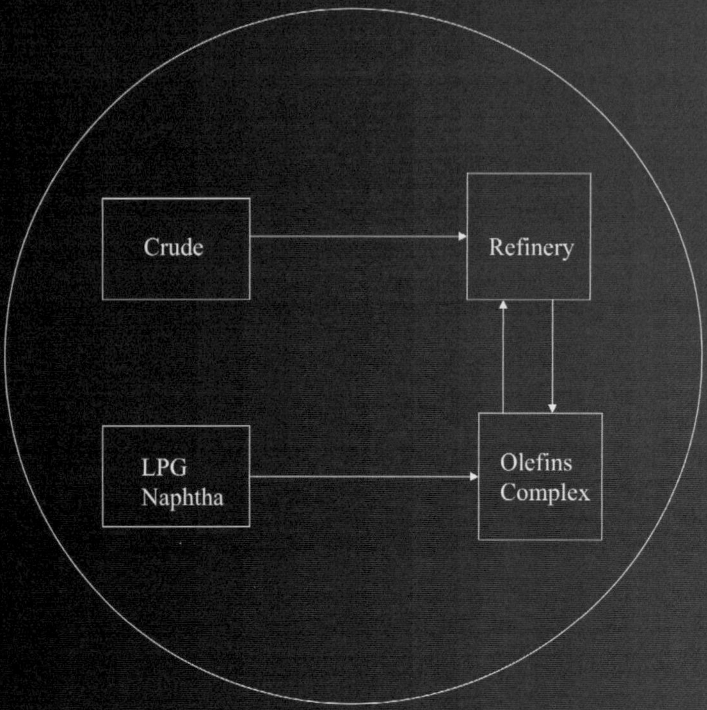

Full benefits requires site-wide planning

Integration Benefits

- Lower storage and shipping cost
- Shared utilities, etc.
- Eliminate middle-man costs
- Increase operating flexibility

Integration Drivers

- Return on capital
- Cost competitiveness
- Secure feedstock supplies and by-product outlets

Source: The Integration of Refining and Petrochemicals, Ron W. Haddoc
President, FINA, Inc, 1999 NPRA International Petrochemical Conferen

Integration Benefits

- Integration improves site return on capital by 2-5% over cycle
- Effective sites capable of greater than 20% refinery runs to petrochemicals

*** Formatting Issue - Best Available Paper/Graphic ***

Opportunities & Challenges in Transportation Fuels Production

Soni O. Oyekan
Marathon Oil & Criterion
&
Sal Torrisi
Criterion
AIChE Regional Process Technology Conference
Galveston, TX
October 2, 2009

An American Company Serving America — National Environmental Performance Track — VPP — Marathon Petroleum Company LLC

Outline

- Crude Oil & Bitumen
  - Availability
  - Upgrading and refining
- Gasoline
  - Low sulfur gasoline
  - Ethanol blending
  - Gasoline benzene
- Diesel
  - ULSD
  - Biodiesel

An American Company Serving America — National Environmental Performance Track — VPP — Marathon Petroleum Company LLC

US Energy Long Term Challenges

- Conventional Crude oil
  - Crude oil quality is declining
  - Increased competition – China & India
  - Oil companies and national oil suppliers relationships are evolving
  - Supply disruptions are more frequent
- Crude oil consumption is expected to increase
- Greenhouse gas emissions and climate issues are ratcheting up

An American Company Serving America

Marathon Petroleum Company LLC

Global Oil Reserves By Country (Source: CAPP)

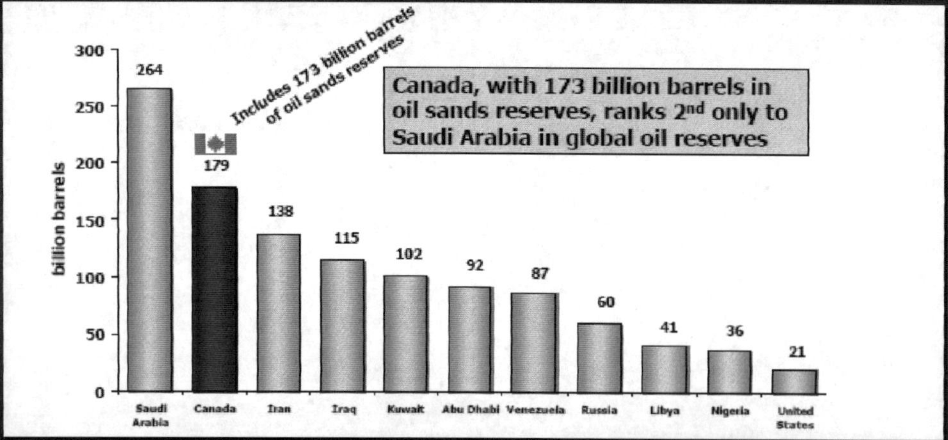

Ref: Emerging technologies for producing paraffins, olefins & aromatics, Michael C. Oballa & Vasily Simanzhenkov

An American Company Serving America

Marathon Petroleum Company LLC

Bitumen Provides Options

Oils Sands: A mixture of sand, crude bitumen, water and clay

Typical Composition

10% bitumen

83% sand

7 % water & clay

- 1 barrel of bitumen is obtainable from 3 to 4 tons of oil sands!
- Bitumen is an extra heavy, sour, poor quality oil
- Significant upgrading required
- 10 % higher GHG generated relative to conventional oil?

An American Company Serving America — Marathon Petroleum Company LLC

Non Facile Commercial Recovery Technologies

Surface Mining

Steam Assisted Recovery

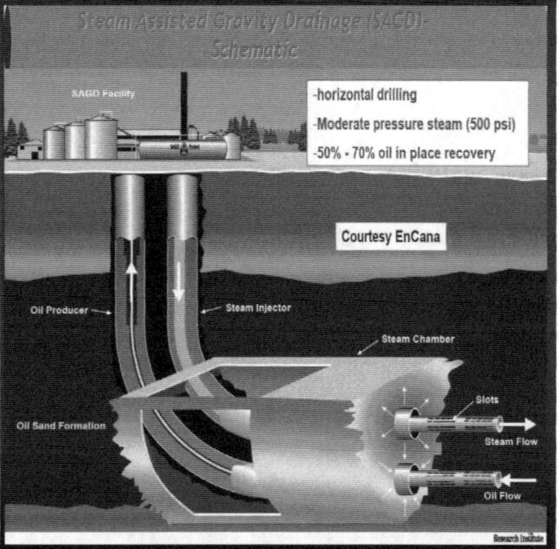

Ref:

An American Company Serving America — Marathon Petroleum Company LLC

Bitumen Pose Significant Challenges

- Bitumen and fractions have higher contaminants
  - Sulfur distributions reflect more refractory compounds
  - Higher refractory nitrogen
  - Higher TAN (Athabasca 3.5 relative to < 1 for conventional oils)
- Significantly higher hydrogen consumption
- Natural gas dependency for fuel and hydrogen
- High requirement for naphtha
- Pipeline infrastructure for market accessibility
- Environmental issues
  - Land reclamation
  - Sour gases and GHG
  - Water conservation

Plant flexibility and operating margins could be impacted.

Bitumen Refining Challenges

Oil Type	Arab Med[1]	Mars	Maya	Hamaca Bitumen[2]	Athabasca Bitumen[3]
Gravity, API	30.7	28.1	20.9	8.4	7.5
Sulfur, wt. %	2.4	2.2	3.4	3.8	4.8
TAN, mg/g	0.25	0.63	0.32	2.8	3.5
Ni, wppm	4	20	52	115	150
V, wppm	18	46	232	388	290

1. Marathon crude oil assay database
2. V/Ni ratio a parameter for palaeo-environmental characteristics…., F. Galarraga, et.al., J. Petro. Science, 61, Issue 1, pages 9-14, April 2008
3. Tutorial on upgrading oil sands bitumen, M. R. Gray, University of Alberta.

An American Company Serving America
National Environmental Performance Track
Marathon Petroleum Company LLC

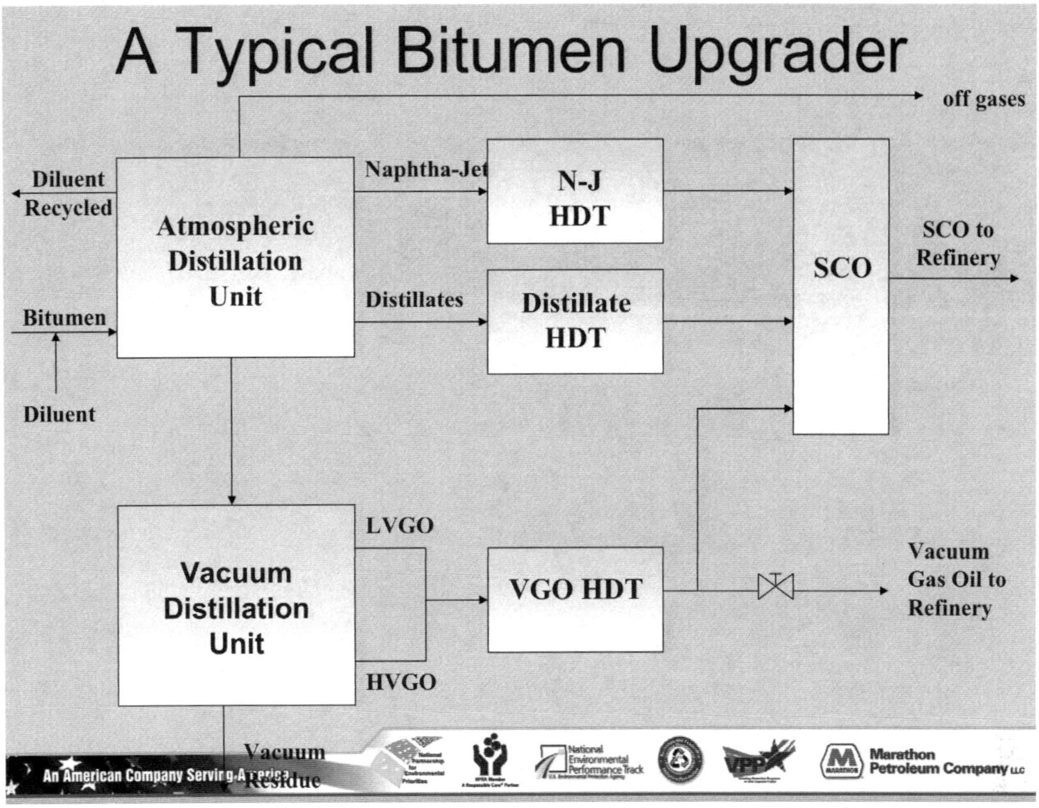
A Typical Bitumen Upgrader
off gases
Diluent
Recycled
Atmospheric Distillation Unit
Bitumen
Diluent
Naphtha-Jet
N-J HDT
Distillates
Distillate HDT
SCO
SCO to Refinery
Vacuum Distillation Unit
LVGO
HVGO
VGO HDT
Vacuum Gas Oil to Refinery
Vacuum Residue
An American Company Serving America
Marathon Petroleum Company LLC

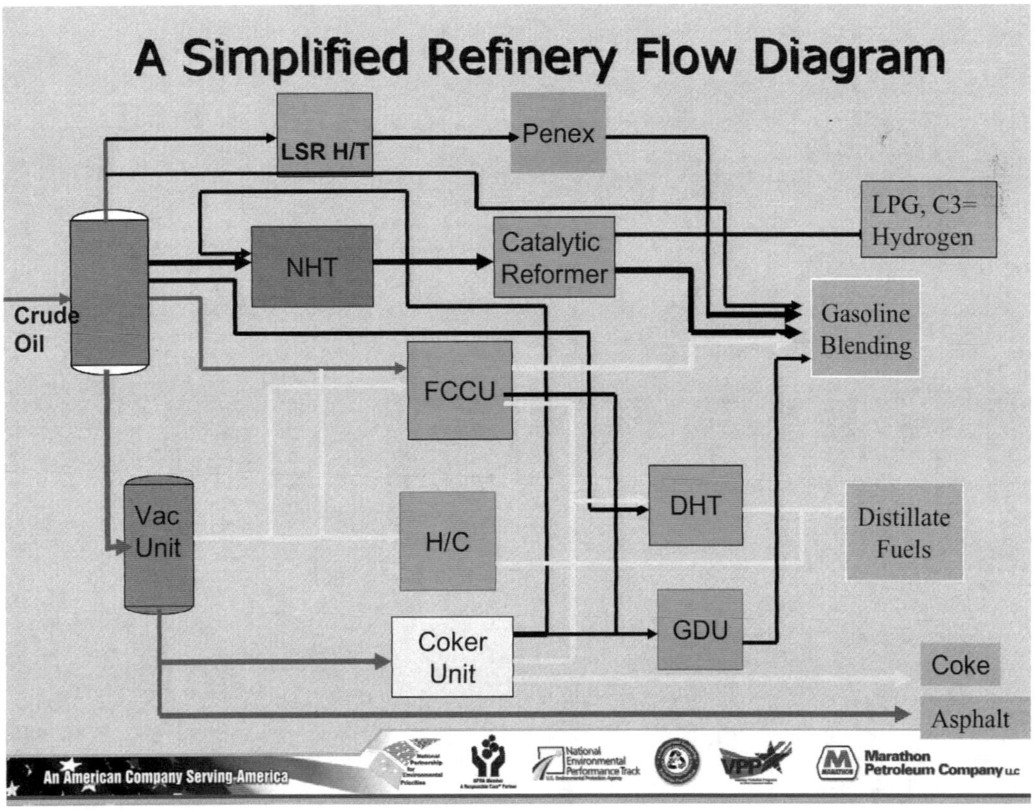
A Simplified Refinery Flow Diagram
LSR H/T
Penex
NHT
Catalytic Reformer
LPG, C3=
Hydrogen
Crude Oil
Gasoline Blending
FCCU
DHT
Vac Unit
H/C
Distillate Fuels
Coker Unit
GDU
Coke
Asphalt
An American Company Serving America
Marathon Petroleum Company LLC

Crack Spread Declining

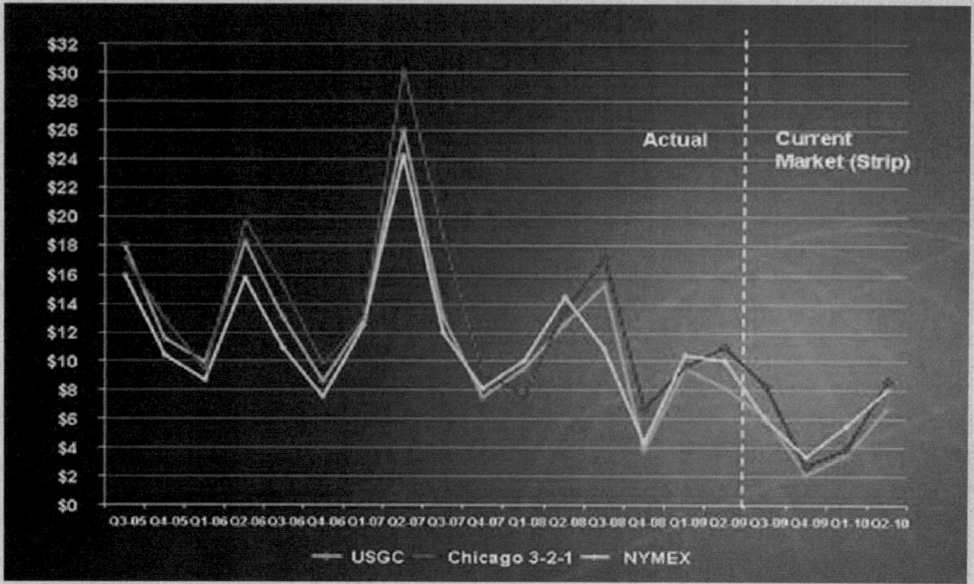

An American Company Serving America — Marathon Petroleum Company LLC

2009 Market Indicators Are Lower

	2008 2Q	2009 2Q
Chicago LLS 6-3-2-1($/BBL)	2.71	**5.73**
Sweet/Sour Diff., ($/BBL)	13.74	3.98
NYMEX WTI ($/BBL)	123.80	59.80

An American Company Serving America — Marathon Petroleum Company LLC

US Refiners Produce Low Sulfur Gasoline

- 2006 gasoline sulfur regulation of < 30 wppm S
  - US refiners met regulation using a variety of technologies and hydrotreating catalysts
- Optimized gasoline desulfurization to preserve octane
- Technologies used included
  - Axens Prime G
  - CD Tech
  - ConocoPhillips Szorb
  - ExxonMobil Scanfining
- Pool gasoline octane losses of 1 to 3 numbers

An American Company Serving America

Marathon Petroleum Company LLC

Ethanol Blending & Naphtha Reforming

- US goal is 35 billion gallons per year of ethanol in gasoline by 2022
- E10 gasoline lead to
  - Increased gasoline volume
  - Gasoline octane giveaway
  - Low severity reformer operations
  - Lower hydrogen production
  - Inefficient catalytic reformer productivity
- Maximize gasoline octane barrels and hydrogen
- Energy savings negligible relative to losses of H_2 and reformate
- Key benefits are less utilization of crude oil for gasoline and enhanced environmental quality

An American Company Serving America

Marathon Petroleum Company LLC

US EPA Gasoline Benzene Regulation

- US EPA Mobile Source Air Toxics II (MSAT II) regulations effective Jan 2011
  - Annual corporate average gasoline pool benzene of 0.62 vol. %
  - Credits and trade arrangements possible
  - By July 2012, maximum refinery average gasoline benzene of 1.3 vol. %
- Catalytic Reformers produce over 80 % of gasoline benzene
- Naphtha Pre-fractionation and Post-fractionation
- Pre-fractionation employs naphtha split to reduce benzene and precursors to the reformer
  - Naphtha Splitter at the crude unit or after NHT
  - Benzene and precursors converted via Bensat and C5/C6 isomerization
  - Achieved ~ 35 to 50 % drop in gasoline benzene at two Marathon refineries
- Gasoline benzene credits option

An American Company Serving America

Marathon Petroleum Company LLC

US EPA Gasoline Benzene Regulation

- Post-fractionation and gasoline benzene minimization processes
  - Extractive distillation for benzene
    - UOP, Uhde
    - Reduced gasoline volume
    - Octane impact
  - Benzene saturation
    - Axens, UOP
    - Gasoline volume maintained
    - Impact on Octane & H2 consumption
  - Benzene alkylation to cumene
- Capital intensive processes for reducing gasoline benzene and maximizing hydrogen production

An American Company Serving America

Marathon Petroleum Company LLC

N. American Clean Diesel Growth Through 2018

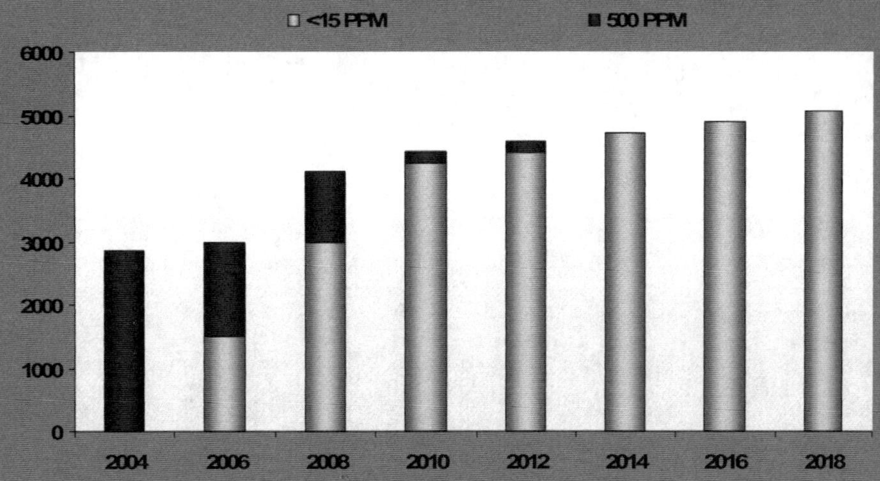

Source: Hart World Refining & Fuels Service

Oil Refiners Met ULSD Regulation

- Technology and oil refining companies honed in on refractory sulfur compounds including 4,6 DM DBT

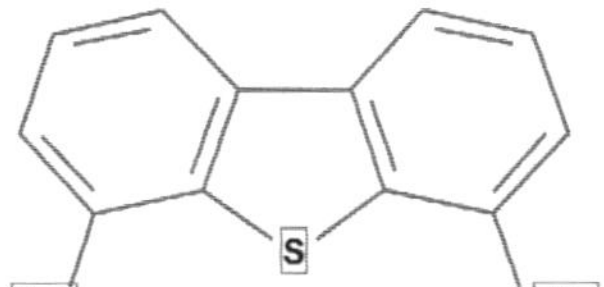

- Utilized a combination of new hydrotreating catalysts, increased reactor volumes, reactor internals process modifications
  - ➢Grassroots Units (~15%)
  - ➢Unit Revamps (~75%)
  - ➢No modifications (~10%)
- Oil refiners achieved < 10 wppm sulfur to meet pipeline requirements

Renewable Diesel Fuels Information

- The mandate for annual renewable fuels use is 36 billion gallons of biofuels by 2022
- Biodiesel is a renewable fuel for diesel engines
  - from natural oils -- soybean oil via reaction with methanol
  - comprised of mono-alkyl esters of long fatty acids
  - meets specifications of ASTM D6751[1]
  - pure biodiesel is B100
- Biodiesel blend is a blend of biodiesel fuel with petroleum based diesel fuel
  - BXX, where XX represents percentage of biodiesel
  - B2, B5 and B20 are used

Ref: Biodiesel Technical Information, ADM Biodiesel Technical Services

An American Company Serving America

Marathon Petroleum Company LLC

US Biodiesel Production On the Rise

- B20 emissions are lower than conventional diesel
  - 21 % lower HCs
  - 11 % lower CO
  - 20 % lower PMs
  - ~ 5 % + NOx
  - 20 % lower PAH

Ref: Clean Cities Fact Sheet US DOE Apr. 2008

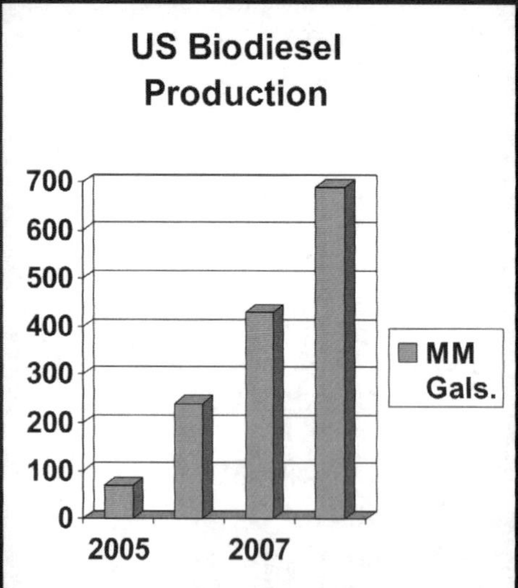

An American Company Serving America

Marathon Petroleum Company LLC

Biofuels Use Will Increase Substantially

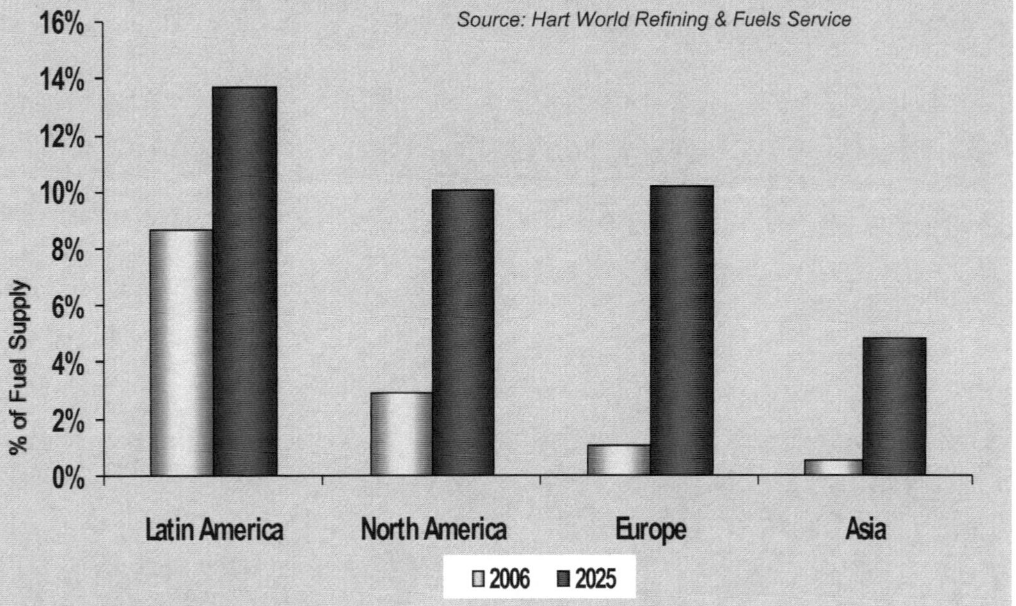

An American Company Serving America

Marathon Petroleum Company LLC

21

Summary

- Unconventional oils will have a significant impact on US refining plants reliability and profitability
- Cost of meeting environmental regulations and GHG re bitumen production will be a factor
- Ethanol gasoline and benzene will challenge US refiners
- Diesel growth is expected worldwide relative to gasoline
- Biofuels use would increase significantly in the next 15 years

An American Company Serving America

Marathon Petroleum Company LLC

The Future -- Gas to Liquids ?

An American Company Serving America

Marathon Petroleum Company LLC

Maximize Diesel Production via Operational or Revamp Options

Mike Siravo

FLUOR®

PURPOSE

- Wide range of options for increasing diesel production
- Low capital options for constrained capital environment
- Revamp options - Study now & implement later

FLUOR®

CRUDE UNIT FRACTIONATION

Maximize lift of LGO from AGO

- Add LGO (Lt. Gas Oil) total draw tray with flow controlled reflux
- Improve stage efficiency with packing
- Install AGO stripper

- **Strip Kerosene product harder to facilitate back-blending into diesel**
- **Add a vacuum column LGO section**

FLUOR®

FCCU MAIN FRACTIONATOR

- **Increase lift of LCO (Lt. Cycle Oil) from HCO/Fractionator Bottoms**
- **Replace trays with packing**
- **Packing reduces column pressure drop**
  - Increases Wet Gas Compressor suction pressure, or
  - Reduces Rx pressure allowing increased cat circulation
- **Note that the higher End Point LCO captured is much tougher to hydrotreat in the ULSD unit vs. Straight Run LGO**
  - Requires evaluation of ULSD unit

FLUOR®

COKER MAIN FRACTIONATOR

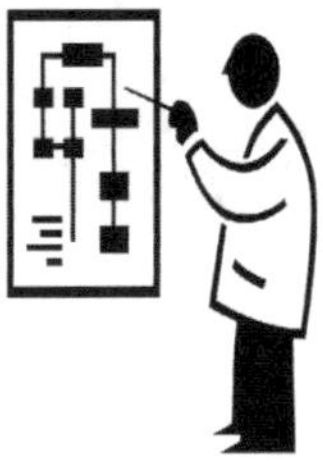

Maximize LGO EP

- **Optimize Pumparounds**
- **Tray debottlenecking**
- **Liquid draw hydraulic debottlenecking**

- Note that:
  - Coker LGO is easier to desulfurize vs FCC LGO
  - Coker LGO has higher cetane vs FCC LGO
  - But Coker LGO has high H2 uptake and heat release in the downstream hydrotreater

FLUOR®

COKER OPERATING CONDITIONS

- **Recycle Coker HGO** and boost refinery margins
  - Convert HGO to LGO and other liquids
  - Very low valued HGO yield decreases
  - Coker HGO consumes valuable capacity in conversion units with low conversion and high operating costs
- **Increase Drum Operating Pressure**
  - LGO yield increases
  - Many new designs - 15 psig
  - Consider increasing to 25 – 40 psig
  - Also reduces compressor HP
- Both have moderate increase in coke and dry gas but liquid yield benefits often predominate

FLUOR®

CATALYST OPTIONS

- <u>FCCU Catalyst:</u> LCO yield and cetane can be increased measurably
- <u>ULSD Catalysts</u>
  - Promotes HDS over HDN with **lower H2 consumption**
  - Unit throughput can be increased if hydrogen limited
  - Optimize with staged catalyst beds allow for increased throughput
- <u>Hydrocracker Catalyst</u>
  - High activity diesel selective catalysts
  - Improved HDN performance
- <u>CFH Catalyst:</u> Increase activity and severity and recover ULSD from product fractionator

FLUOR®

DEBOTTLENECKING STUDIES

- Many ideas presented so far may require debottlenecking for implementation
- Methodology:
  - Start with representative operating data from multiple time periods
  - Develop initial concepts with open art information
  - Progress at later stages with licensor or specific supplier input

FLUOR®

INDIVIDUAL UNIT vs. REFINERY-WIDE STUDIES

Study considerations:

- Optimize Unit? - May sub-optimize the refinery
- Multi-unit optimization may be acceptable (such as FCC/HCU system)
- Refinery-wide LP study - Capture all impacts

FLUOR®

Maximize Diesel Production

Case Studies

FLUOR®

NO COST DEBOTTLENECKING

- <u>Optimized hydrotreater recycle compressor</u>
  - Centrifugal recycle gas compressor was not at max speed
  - Recycle gas molecular wt. was optimized
    - ↗ H_2 purity was too high
  - H_2 partial pressure at Rx outlet increased
  - Catalyst deactivation rate was reduced significantly

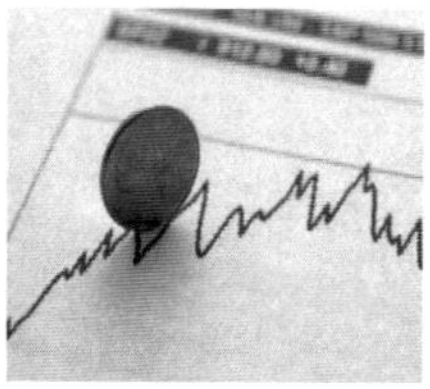

FLUOR®

LOW COST DEBOTTLENECKING

- <u>Feed manifold bottlenecks removed</u>
  - Refinery had three distillate hydrotreaters
  - Feed manifold was old and complex
  - Line size increases yielded <u>capacity increase</u>
- <u>Refinery with multiple distillate hydrotreaters</u>
  - Multiple distillate feed streams
  - Installed feed manifold piping and controls to <u>optimally</u> pool the components
  - Effectively allocated feed quality to each unit
  - <u>Maximized combined capacity</u>
  - Optimized catalyst run lengths

FLUOR®

LOW COST OPTIMIZATION

- CFH Product Stripper Revamp
  - Modified to add diesel side draw
  - Diesel was further treated in distillate hydrotreater
  - Produced more diesel
  - Also created room in FCCU that was then filled with purchased gas oil feed
  - FCCU made more diesel and gasoline from feed quality improvement

FLUOR®

MODERATE COST REVAMP

- **Converted Small/Low Severity CFH to ULSD**
  - **20 MBPD, 1100 psig** hydrotreater was being used to reduce metals and nitrogen from only 20% of the feed to an FCCU complex
  - Operating cost was high vs. FCCU margin impact
  - The unit was converted at low capital cost to produce 20 MBPD of ULSD
  - Feed was straight run LGO
  - **Rate of return was above 40%**

FLUOR®

CONVERT RECYCLE HCU TO ONCE THROUGH

- A recycle HCU ran 95% conversion to naphtha & lighter
- Recycle was mostly recoverable high cetane diesel
- Revamp eliminated recycle & added diesel recovery column
- Recycle was replaced with fresh feed
- Naphtha & lighter production was mostly unchanged, but
- **Fresh Feed capacity increased 35%**
- **15 MBPD high cetane diesel was produced**

FLUOR®

FLUOR CAPABILITY

- Intense focus on **creative** but **practical** margin improvement concepts
- Full range of study services: Operating data analysis, concept and scope development, equipment rating, spreadsheet modeling to complex refinery LP modeling, cost estimating
- Open art basis - Avoid tunnel vision introduced by those who license technology
- Studies tailored to capital availability
- Wide understanding of industry best practices

FLUOR®

CONCLUSION

- There are many paths to increasing diesel regardless of budget
- Consider partnering with someone that has the capability to help you meet your business goals

FLUOR®

QUESTIONS? ? ?

Mike Siravo

Fluor, Sugar Land

281-263-8460

FLUOR®

AUTHOR INDEX